# 30-SECOND
# LEONARDO
# DA VINCI

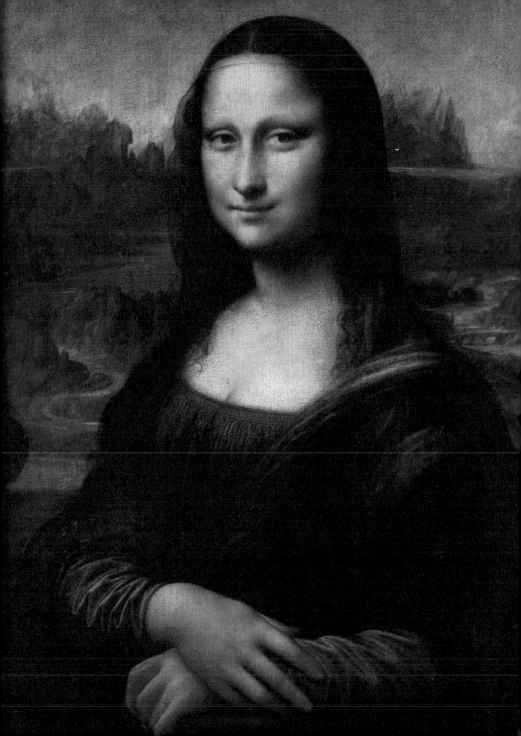

# 30-SECOND
# LEONARDO
# DA VINCI

THE 50 KEY ASPECTS OF HIS WORKS, LIFE & LEGACY,
EACH EXPLAINED IN HALF A MINUTE

Editor
**Marina Wallace**

Foreword
**Martin Kemp**

Contributors
**Francis Ames-Lewis**
**Juliana Barone**
**Paul Calter**
**Brian Clegg**
**Matthew Landrus**
**Domenico Laurenza**
**Marina Wallace**

Illustrations
**Ivan Hissey**

**IVY PRESS**

This edition published in the UK in 2018 by
**Ivy Press**
An imprint of The Quarto Group
The Old Brewery, 6 Blundell Street
London N7 9BH, United Kingdom
**T** (0)20 7700 6700 **F** (0)20 7700 8066
www.QuartoKnows.com

First published in hardback in 2014

British Library Cataloguing-in-
Publication Data
A catalogue record for this
book is available from the
British Library

ISBN: 978-1-78240-582-5

This book was conceived, designed
and produced by
**Ivy Press**
58 West Street, Brighton BN1 2RA, UK

Creative Director  **Peter Bridgewater**
Publisher  **Jason Hook**
Editorial Director  **Caroline Earle**
Art Director  **Michael Whitehead**
Commissioning Editor  **Kate Shanahan**
Project Editor  **Jamie Pumfrey**
Designer  **Ginny Zeal**
Glossaries Text  **François Gaignet**

Printed in China

10 9 8 7 6 5 4 3

# CONTENTS

# FOREWORD
## by Martin Kemp

Leonardo roundly criticized the "abbreviators"— those who try to take shortcuts to knowledge. This book is necessarily about abbreviation. We have, in the face of the daunting range of delightful and often exceedingly difficult material in his surviving legacy, to effect some kind of summary and synthesis. We could easily become lost in his diversity. Scholars have undertaken substantial studies of Leonardo as an artist and theorist, as an architect, as a stage designer, as an engineer, as an anatomist, a mathematician, physicist, as a geographer and geologist, and many more things. However, as Marina Wallace stresses, there is a unity at the heart of his diverse pursuits. Everywhere he sought the core of shared "science" that granted validity to his intellectual and practical activities. "Science" in this sense means a systematic body of knowledge that is verified by "experience;" that is to say that is drawn out of the divine order of nature.

Practical pursuits, previously regarded as crafts (including painting), were being endowed in the Renaissance with their own theories, and Leonardo was in the forefront of promoting "sciences" that combine abstract mathematics and the material realities of inventing things. Whatever he was doing, he was "remaking" nature on his own terms but always on the basis of natural law.

You are encouraged to participate in this inner unity by accessing the book in a nonlinear manner. Leonardo was the lateral thinker of lateral thinkers. When thinking about geometry, he would by association think about the motion of the water; thinking about turbulent water he would think about the curling of hair; thinking about curly hair he would think about the vortices of blood in the heart; thinking about blood in the heart he would think about vortices in mighty storms or deluges. You can make your own cross-connections. It is an exciting journey.

**The portrait of a man in red chalk**
*Although this is not acknowledged as a portrait of Leonardo, it seems to sum up the main characteristics of the man and artist.*

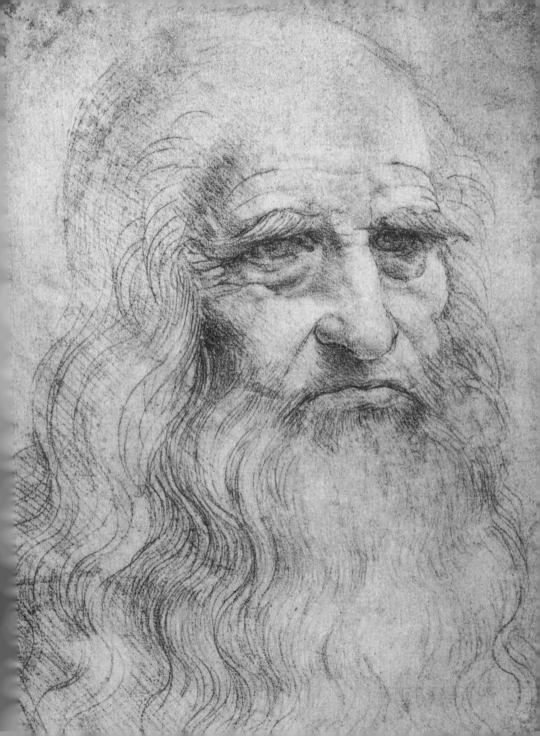

# INTRODUCTION
by Marina Wallace

### Leonardo da Vinci

Leonardo was born in Vinci, a small town near Florence, on April 15, 1452. His father, Ser Piero di Antonio da Vinci, was a notary, his mother, Caterina, a farmer's daughter. In 1472, Leonardo's name appears in the account book of the Florentine confraternity Compagnia di San Luca, and in 1476 he was in the studio of the painter and sculptor Andrea del Verrocchio (1435–1488).

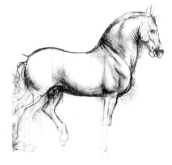

Between 1478 and 1492, Leonardo was involved in a series of different endeavors: a commission for an altarpiece for a chapel in the Palazzo Vecchio in Florence; one for the domed crossing, or tiburio, of Milan cathedral; one for the design of an equestrian statue for Ludovico Sforza; another for an altarpiece depicting the Virgin of the Rocks for the Milanese Confraternity of the Immaculate Conception; and the stage set for a *Feast of Paradise* to be performed in the Castello Sforzesco in Milan.

During this period he also traveled to Lake Como in Lombardy, visiting Valtellina, Valsassina, Bellagio, and Ivrea. In 1489, he is also recorded to have inscribed one of his skull studies for his book *On the Human Figure* (now at the Royal Library, Windsor, UK). Such summary account of only 14 years of Leonardo's life provides a sample of the variety of activities that he typically engaged with, moving from one medium to another: from painting to sculpture, to planning canals for governing the flow of water in cities, to designs for war machines to be supplied to his patrons, to inventing tools and mechanical objects, and relentlessly taking notes on what he observed in nature and all around him.

He traveled across several geographical locations, mainly between Florence and Milan, but also to Mantua, Bologna, Urbino, Cesena, Pesaro, Rimini, and Piombino. Finally he moved to France in 1516, accompanied by his pupil Francesco Melzi and his assistant Salai, where he served the French king Francis I. Throughout his life he received payments and gifts from his patrons for his work, which was often delivered late. His pension, granted by Francis I, was a highly respectable sum of money: 2,000 ecus soleil for two years. His assistants also received a good royal payment: Melzi 800 and Salai 100 ecus soleil.

On May 2, 1519, Leonardo died in the manor house of Clos Lucé. The fame he had gained during his lifetime heightened after his death. His name is now known universally. One of his paintings—the *Mona Lisa*—is the most famous in the world; another—the *Last Supper*—perhaps the second most famous. The *Vitruvian Man* is the most famous drawing ever. The sheer quantity of his writings is daunting, the quality astonishing: the number of surviving pages amounts to roughly 6,000 and would fill perhaps 20 volumes of a modern book. His handwriting is difficult to read as it unfolds from right to left in mirror script (he was left-handed).

## "Renaissance Man"

What is apparent in Leonardo's prodigious output of paintings, drawings, and manuscripts is that he was relentless in his observation of nature, be it the human body or the movement of water in valleys and mountains. Any analysis we may carry out is always partial and needs to take into account a much wider context. We must consider that all of Leonardo's investigations, his observations and records in his notebooks, have, at their base, deep and connected thoughts. Leonardo strove to ascertain principles, causes, and reasons, and to verify his conclusions through experimentation and case studies. His method relied on experience. Leonardo embodied skills and knowledge that would nowadays require dozens of different specializations, from geology to anatomy, physics, music, engineering, painting, and many more. The term "Renaissance Man" indicates the ability of one individual to embrace many different disciplines. Leonardo was the "Renaissance Man" par excellence.

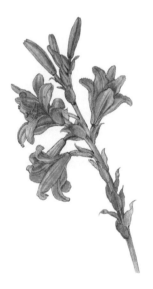

## Artist and Scientist

Leonardo's firm belief in the interconnectedness of all natural forms and the all-governing creative power of nature is manifest in his observations and experimentations. One has the feeling that nothing passed by Leonardo's eye unnoticed. His mind was always at work trying to unravel the deepest causes of all phenomena, aiming to simplify complex principles. His ability to visualize was extraordinary. His thought process gained from his outstanding skill in drawing detailed depictions of what he imagined and saw. His sketches and annotations should be seen as brainstorming on paper. His scientific mind, always in search of ways in which he could prove or disprove his hypotheses, was at work also in his artistic output. His paintings and drawings, of astounding beauty, are part of the whole enterprise, not to be considered separately from his scientific endeavors.

## How the Book Works

The approach in *30-Second Leonardo da Vinci* is to zoom into Leonardo's all-encompassing mind to paint a complete picture out of individual details. The book is organized into seven chapters. Each of these addresses a major area of Leonardo's scientific and artistic activity. The full picture will hopefully give a sense of how a typical "Renaissance Man" functioned, and how many different fields he dipped his mental brush in. Each entry is made up of a **30-second theory**, to highlight and focus on a particular detail, a fragment contributing to sketching out the whole picture. This is further distilled into a **3-second sketch**. A **3-minute masterpiece** is included against each entry to bring to the foreground an example of a picture, sculpture, or object that best illustrates the ideas discussed in the entry. The **3-second biographies** prime the reader for a fuller landscape, providing details of key figures who influenced, or were influenced by, Leonardo. Feature spreads included in chapter each focus on a single example of Leonardo's artistic genius.

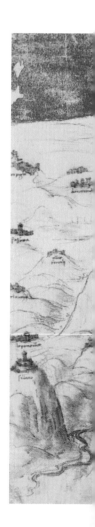

## Topographer and cartographer

*This bird's-eye map of Tuscany, from around 1503, is just one*
*of the many examples of Leonardo's incredible dedication*
*and artistic accuracy. Pinpointing locations and defining*
*landscapes is not only an astonishly beautiful art piece but*
*also a technical achievement unmatched of the time.*

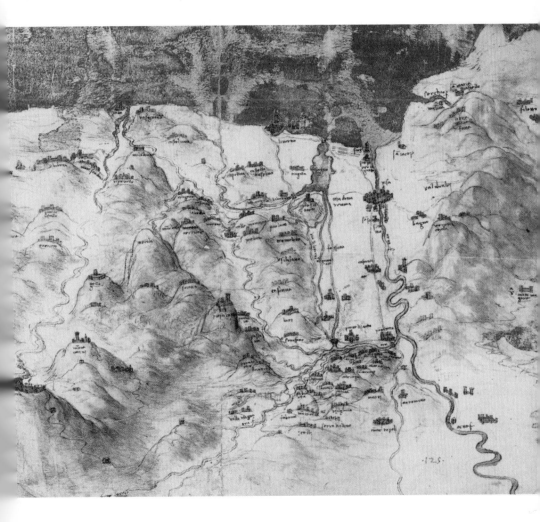

# PAINTING & SCULPTURE

# PAINTING & SCULPTURE
## GLOSSARY

**academic art** After the acceptance of the visual arts as liberal arts, thanks to the efforts of Leonardo, the late 16th century saw the creation of art schools, or art academies, throughout Europe that were to last until the 19th century. Sponsored by influential patrons of the arts, their aim was to educate young artists in the manner of the classical theories of art of the Renaissance, and these rules and conventions came to be known as academic art.

**aerial perspective** The visual effect of moisture and dust in the atmosphere on landscapes, particularly noticeable on distant objects, where colors seem to become bluer and tones more muted—the farther the object, the more its color matches that of the surrounding atmosphere. In painting, this gives an impression of depth and distance and is achieved by using true colors for foreground objects, gradually muting both colors and tones, as well as decreasing outline sharpness for the depiction of objects in the background. Also called atmospheric perspective.

**chiaroscuro** An Italian term, meaning "light–dark," that describes the effect of strongly contrasting light and shade on a work of art. The technique was pioneered by Leonardo as an effective way of creating the illusion of depth and volume in a painting. In the early 17th century, Caravaggio and Rembrandt used the same technique to add a dramatic dimension to a scene.

**liberal arts** In the Classical world, the seven liberal arts were the studies worthy of a free man (*homo liber*) and were divided into *trivium* (grammar, rhetoric, and logic) and *quadrivium* (arithmetic, geometry, music, and astronomy). The lowly position of the visual arts as "vulgar" became increasingly contested during the early Renaissance, and efforts were made to raise them from the status of manual skills to that of liberal arts. Leonardo, more than anyone else, was responsible for promoting the idea of the painter as a creative thinker and, by 1500, both painting and sculpture were acknowledged as liberal arts.

**paragone** Italian word meaning "comparison" (of the arts), it refers to a debate during the Renaissance regarding the merits of painting versus those of sculpture and which could reproduce the various forms of nature most successfully. In the *Treatise on Painting*, Leonardo argues for the supremacy of painting over the arts of sculpture, music, and poetry.

**sfumato** From the Italian word *fumo*, meaning "smoke," the term describes the very subtle blending of tones or colors into each other without noticeable transition. Leonardo described this technique as a gradation "without lines or borders, in the manner of smoke." Recent X-ray analysis of the *Mona Lisa* painting showed that Leonardo used up to 20 layers of paint as thin as two microns each to achieve this effect.

**Treatise on Painting** Throughout his life, Leonardo wrote a large number of notes and manuscripts, in his famous right-to-left, mirror-image script, and intended to publish treatises on different subjects. Instead, they were left to his friend and pupil Francesco Melzi, who started to collate them into the *Libro di Pittura*. Published in French and Italian in 1645 as *Trattato della Pittura*, and translated into English in 1817, the *Treatise on Painting* has been described as "the most important document in the entire history of art." It begins with careful instructions on drawing the main features of human anatomy, then moves on to techniques of rendering motion and perspective. Good composition, inventiveness, the expression of emotions, the effects of light, shadow, and color, and many other subtle points of artistic expression are discussed, with the importance of meticulous study of the subject emphasized throughout.

# THE TREATISE ON PAINTING

## the 30-second theory

Leonardo's *Treatise on Painting* became the main vehicle for the transmission of his theoretical ideas until the 19th century. Although no finished treatise by Leonardo himself survives—and what we now know as the *Treatise* is a 16th-century compilation by his pupil, Francesco Melzi—the materials it includes were copied directly from his notebooks and are therefore faithful to his teachings. It starts with a discussion (*paragone*) on the primacy of painting over sculpture, poetry, and music, in which the principles of what he considers the science of painting are stated. It then provides a series of precepts to painters, stressing the importance of nature as the primary guide. Next it discusses the human body, with a focus on proportion, motion, and expression, touching on the basic tenets for narrative paintings and ending with observations on drapery. Finally, it follows lengthy scientific discussions on light and shadow. These not only reveal Leonardo's ideas about color and aerial perspective, but include instructions on how to depict trees, clouds, and the horizon. The *Treatise* shows Leonardo's extraordinary mind at work, in his quest to understand the "laws" of nature as the founding principles for the science of painting, and to instruct the painter in how to re-create the natural world according to those principles.

**3-SECOND BIOGRAPHIES**
FRANCESCO MELZI
ca. 1492–ca. 1570
Italian painter and heir to Leonardo's manuscripts.

NICOLAS POUSSIN
1594–1665
French painter, who lived in Rome most of his active life.

**30-SECOND TEXT**
Juliana Barone

*Exceptional in the breadth of analysis and skill as a draftsman, Leonardo's notebooks contain several studies of the movement of the human figure.*

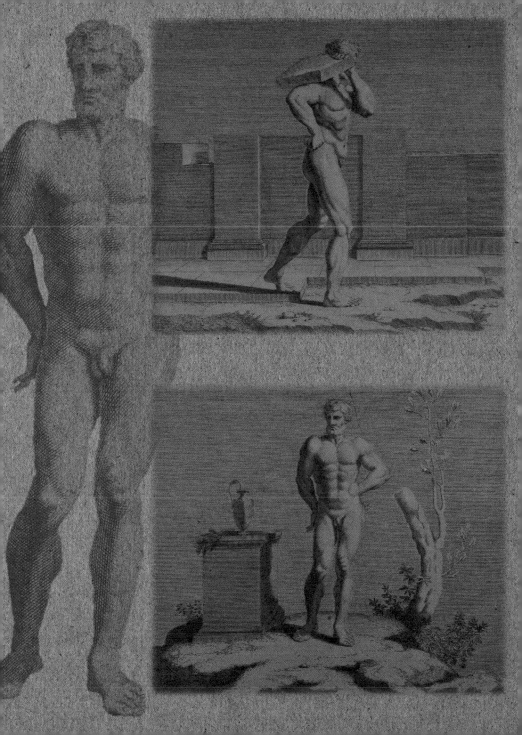

# PARAGONE

## the 30-second theory

### 3-SECOND SKETCH
In his *Treatise on Painting,* Leonardo drew comparisons between the arts, and aligned painting with poetry, seeking to show that painting should be included among the liberal arts, the humanities.

### 3-MINUTE MASTERPIECE
*Paragone* led to frequent intellectual quarrel in Leonardo's day. It was stimulated by the Roman poet Horace's aphorism *"ut pictura poesis"* (as a painting, so a poem). Views on the superiority of painting over sculpture are expounded in a debate in Castiglione's *The Book of the Courtier,* published in 1528. The issue was farther debated in a series of lectures given by Benedetto Varchi in 1546, in which Varchi maintained that painting and poetry had imitation in common, but that sculpture was superior to painting.

**"Painting laments that it has** been excluded from the liberal arts," concluded Leonardo in the section of his *Treatise on Painting* that deals with the *paragone,* or comparison of the arts. Comparing and contrasting painting with poetry and music, he sought to show that painting does indeed deserve to be numbered among the liberal arts. Painting, Leonardo argued, is superior to poetry because it appeals to the eye not to the ear. The eye—the "window of the soul"—is the foremost sensory organ, because it is through the eye that we can best comprehend the limitless character of nature's works. Leonardo proposed that painting affects one's emotions more quickly than poetry because a painting can be seen and understood instantaneously, while a poem takes time to read. Similarly, painting "is superior in rank to music, because it does not perish immediately after its creation, as happens with unfortunate music." Leonardo also considered painting as worthier than sculpture—it requires intellectual activity on the part of the artist. Moreover, unlike the sculptor, the painter can show all the aspects, and all the colors, of nature.

### RELATED RESEARCH
See also
THE TREATISE ON PAINTING
page 16

CHIAROSCURO & SFUMATO
page 22

NATURE
page 26

### 3-SECOND BIOGRAPHIES
BALDASSARE CASTIGLIONE
1478–1529
Writer, diplomat, and courtier.

BENEDETTO VARCHI
1503–1565
Florentine historian, poet, and critic.

### 30-SECOND TEXT
Francis Ames-Lewis

*Sculpture, such as Bartolomeo Amannatti's Leda and the Swan, Leonardo considered, required mere physical effort, whereas painting was more than a mechanical process; for him it was the highest of the arts.*

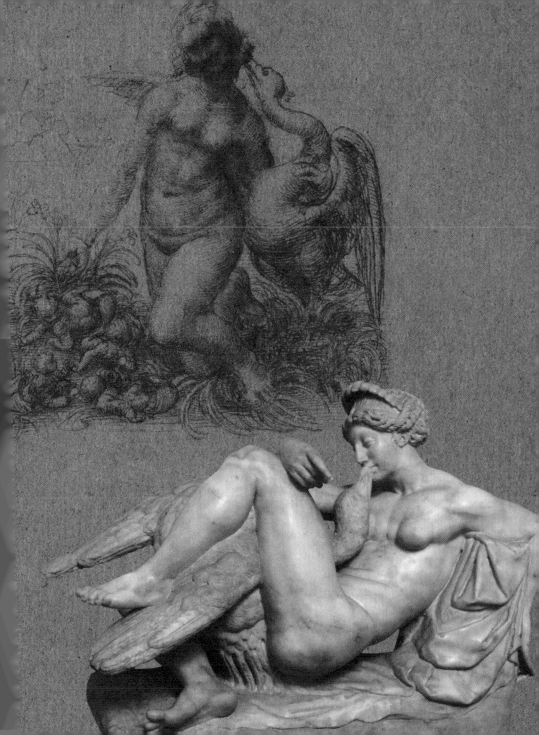

# DECORUM OF MOVEMENT & GESTURE

## the 30-second theory

### A significant section of Leonardo's

*Treatise on Painting* is taken up with advice on representing movement, and on ensuring the appropriateness of movements and gestures to the figure shown. "The actions of men are to be expressed in keeping with their age and station," he wrote. An old man should move sluggishly; young women should be demure, with lowered eyes; an old woman should be shown as "shrewlike and eager, with irascible movements." Children must be shown "with lively, wriggling actions when they are seated." In drawings for the unfinished *Adoration of the Magi*, Leonardo studied a range of figure poses and gestures to suit the different ages of the three magi. A figure's movement must also reflect his mental state, whether he is angry, tearful, happy, or frightened; and hand gestures should mirror his emotional response. The ideal exemplar of this advice in Leonardo's own work is his mural painting in Milan of the *Last Supper*. Each of the apostles reacts differently to Christ's words, and their movements, hand gestures, and facial expressions aim to portray the nature of the emotions they feel.

**3-SECOND SKETCH**
A painted figure's movements and gestures need to be appropriate to its age, gender, and station in life, and to express its emotional state.

**3-MINUTE MASTERPIECE**
Later painters adopted and reinforced Leonardo's ideas on the decorum of movement and expression. Soon after Leonardo's *Treatise* was published in 1651, Charles Le Brun, court painter to King Louis XIV, wrote a treatise entitled *Méthode pour Apprendre à Dessiner les Passions*, published posthumously in 1698. Following Nicolas Poussin's theoretical ideas on facial expression, Le Brun sought to codify the visual representation of human emotions. This treatise continued to be important for "academic" painters until the late 19th century.

**RELATED RESEARCH**
See also
THE TREATISE ON PAINTING
page 16

THE LAST SUPPER
page 24

**3-SECOND BIOGRAPHIES**
NICOLAS POUSSIN
1594–1665
French painter who worked principally in Rome.

CHARLES LE BRUN
1619–1690
French "academic" painter and artist-politician.

**30-SECOND TEXT**
Francis Ames-Lewis

*Old and young men sketched in the streets of Milan became Leonardo's apostles. Their mannerisms, gestures, and exchanges showing their responses to Christ's words and actions at the Last Supper.*

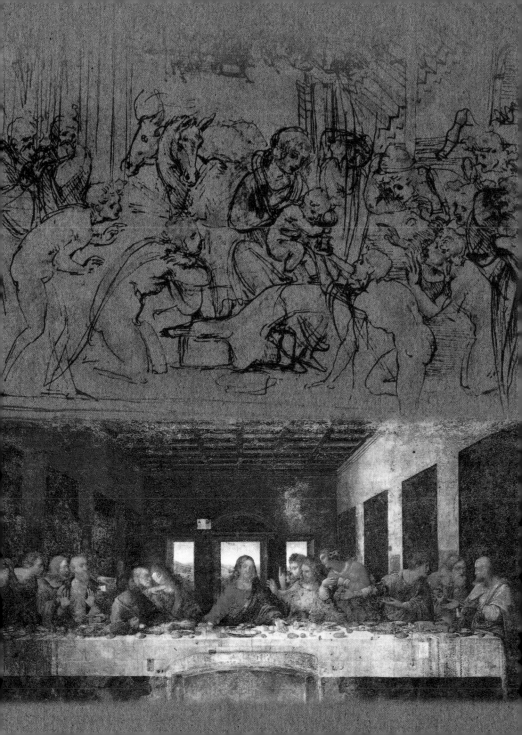

# CHIAROSCURO & SFUMATO

## the 30-second theory

Chiaroscuro—literally, "clear-dark"—describes the transitions from lighted areas to dark shadows in paintings, aiming to give the illusion of three-dimensional forms. Sfumato, or "smoky," describes the particularly sophisticated depiction of the passage from light to dark developed by Leonardo, who wrote that light and shade should merge "without lines or borders, in the manner of smoke." Galileo later wrote that sfumato was a technique of blending one tone with the next without crudeness, "by which paintings emerge soft and round, with force and relief." In optical studies of the 1490s, Leonardo analyzed extensively the tonal transitions generated by the fall of light on objects. This experimental work lay behind the highly evolved sfumato of the heads of figures in the London *Virgin of the Rocks*. Also fundamentally important was Leonardo's sophisticated handling of the oil technique, which 15th-century Italian painters had only recently mastered. Enabled by the use of oil paint, subtly atmospheric sfumato effects using imperceptible transitions from highlight to shadow can be observed in the head of St. Anne in the *Virgin and Child with St. Anne*, and in the study in black chalk Leonardo made for it. The soft tonal gradients reinforce St. Anne's gentle, sensitive expression as she looks lovingly down toward the Madonna.

**30-SECOND TEXT**
Francis Ames-Lewis

*Through sophisticated handling of light, shade, and color, Leonardo achieved a subtle realism that describes three-dimensional form and emotional expression.*

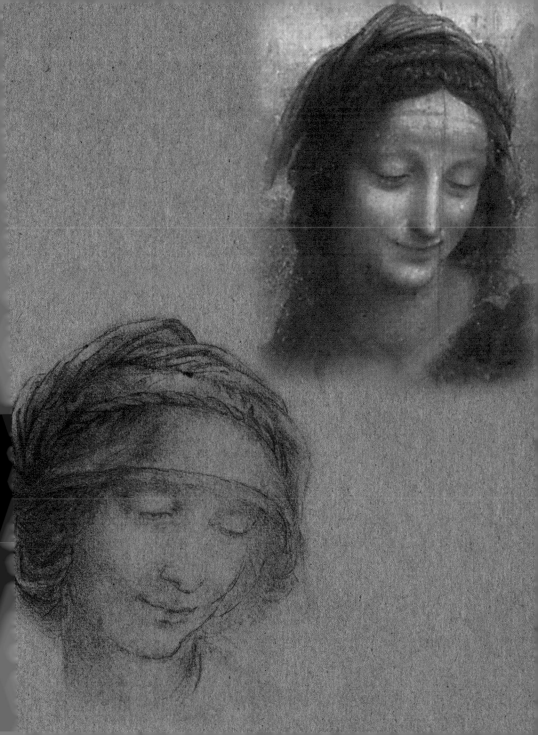

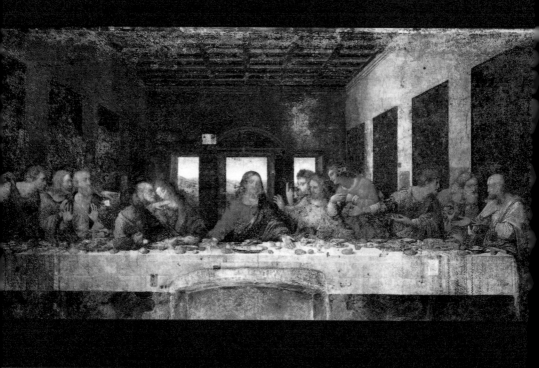

# THE LAST SUPPER

The *Last Supper* was executed as part of Ludovico Sforza's program for the enlargement and redecoration of the Dominican complex of Santa Maria delle Grazie in Milan, and covers a wall of the refectory. Dating from ca. 1494–98, it is one of the most exalted creations of narrative painting. It represents the moment of Christ's shocking announcement that he is to be betrayed by one of his apostles, which is followed by the Institution of the Eucharist as evoked in his gesture of reaching for a glass of wine and offering of bread.

Leonardo was revolutionary in both expressing and orchestrating the reactions of the apostles, who respond individually (as their gestures and facial expressions reveal their inner characters) and as part of groups of threes (which develops intermediate levels of narrative and compositional interest). In the group to Christ's right, Judas is shown suddenly leaning backward and seizing his bag of money; Peter is moving forward in disbelief and holding a knife; and John is conveyed in all his tender innocence. As Leonardo states in his .*Treatise on Painting*, "The figure is most praiseworthy which best expresses, by its action, the passion of its soul." Leonardo's successful representation of the scene is also intrinsically linked to his skillful construction of space as a visual extension of the refectory. Making use of perspective, with the vanishing point on Christ's head, the construction is visually compelling, but there are signs that Leonardo exercised artistic license to accommodate the narrative. The scene is not represented from the actual point of view of someone in the refectory (only the underside of the table would have been visible), but from an ideal position more than twice as high. The size of the figures has also been increased (they cannot all sit behind the table) in order to explore expression and meaning at its highest level.

From the moment it was unveiled, the painting enjoyed enormous appreciation in Italy and beyond, influencing artists such as Rubens and Rembrandt, and playing a crucial role in French academic art theory. Departing from the conventional fresco, the experimental technique Leonardo adopted (egg tempera and oil glazes in some places) allowed him not only to revise but also to create more subtle effects of light and modeling, however, as a result, the paint surface soon deteriorated and has since undergone several restorations.

*Juliana Barone*

# NATURE

## the 30-second theory

In his *Treatise*, Leonardo called painting the "daughter of nature," because all things that nature produced can be found within painting. Returning to his *paragone* arguments, he claimed that painting is superior to poetry because the painter depicts nature's works more truly than the poet does. Moreover, painting is superior also to sculpture because it uses color and aerial perspective. In addition, it can depict clouds, storms, night scenes, grasses and flowers, landscapes, and much more. The painter's primary role is the exact imitation of nature through firsthand, empirical observation. His experience of natural laws is critical, whether he is studying the myriad movements of a human figure, the proportions of a horse, the growth and branching of trees, the fall of light and shadow, or the effects of distance on details, colors, and tonality of objects. The painter must not seek to improve on nature, nor show only the beautiful in nature. Rather, he should record nature's full variety, not merely copying nature slavishly, but exploiting his inventiveness (*invenzione*). Finally, the painter should not rely on memory nor imitate other painters but should observe nature, study its laws, and build with imagination on this foundation.

**3-SECOND SKETCH**
For Leonardo, nature is the mother of painting, and the painter must base his creative inventiveness on the exact imitation of nature.

**3-MINUTE MASTERPIECE**
Leonardo's insistence on the authority of nature was not new. In his *Libro dell'Arte* (ca. 1400), Cennino Cennini wrote of copying from nature as "the most perfect steersman that you can have." Leon Battista Alberti wrote in his *De Pictura* (1435) that the painter should both follow nature in his way of life, and choose what is most typical and best in nature when creating a work that is more beautiful than nature itself.

**RELATED RESEARCH**
See also
THE TREATISE ON PAINTING
page 16

PARAGONE
page 18

**3-SECOND BIOGRAPHIES**
CENNINO CENNINI
ca. 1370—ca. 1440
Florentine painter and writer of the technical treatise *Il Libro dell'Arte* (*The Craftsman's Handbook*).

LEON BATTISTA ALBERTI
1404–1472
Florentine humanist and polymath, author of *De Pictura* (*On Painting*).

**30-SECOND TEXT**
Francis Ames-Lewis

*Disciplined in both art and scientific study, Leonardo based his work on meticulous observation of nature, his topics of interest including botany, water, aerodynamics, anatomy, and geology.*

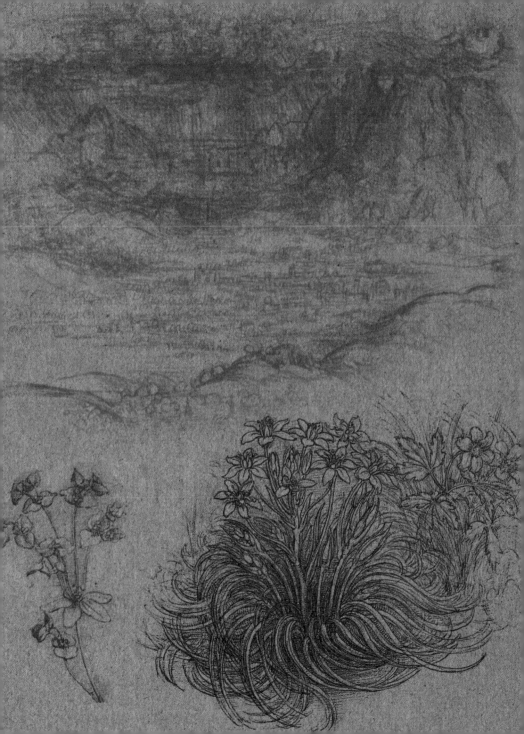

# PORTRAITURE

## the 30-second theory

**3-SECOND SKETCH**
In his portraits, Leonardo explores unusual and dynamic poses, and seeks to represent personality by capturing fleeting and engaging facial expression.

**3-MINUTE MASTERPIECE**
Perhaps the first painter to learn from Leonardo's originality and to absorb it into his practice was Raphael: his *Maddalena Doni* was closely based on the *Mona Lisa*. Through his work, Leonardo's achievements in his portraiture, in activating his sitters' poses and expressions and in defining their emotional states, became more widely known. Later painters who were much indebted to Leonardo's innovative contributions to the art of portraiture include Rembrandt, Velázquez, and Ingres.

**Leonardo painted only a few** portraits, but they are all deeply original, and his portraiture had a profound influence on the development of the genre. Only one shows a male sitter, the *Portrait of a Musician*. His drawing of *Isabella d'Este, Marchioness of Mantua* shows the figure in three-quarter view while she twists her head into profile. This was conventional for a sitter of the highest rank, but it results here in a remote, idealized image. Leonardo never made the promised painted version, which suggests that he was dissatisfied with this format. In his portraits of Florentine bourgeois women, *Ginevra de' Benci* (the painting here) and *Mona Lisa*, the sitter is shown full-face. The first engages the viewer with an unnerving, indecorously piercing gaze, which probably derives from Flemish prototypes. His two portraits of mistresses of Ludovico Sforza, Duke of Milan—*Cecilia Gallerani* and *Lucrezia Crivelli*—show each sitter in a state of transitory movement, both in her pose and, more significantly, her emotional state. These portraits are paradigms of Leonardo's constant search to find ideal ways of representing what he called "the movements of the mind."

**RELATED RESEARCH**
See also
PARAGONE
page 18

MONA LISA
page 110

**3-SECOND BIOGRAPHIES**
RAPHAEL
1483–1520
Painter and architect born in Urbino, who worked in Florence and especially in Rome.

DIEGO VELÁZQUEZ
1599–1660
Spanish painter who worked principally for the court of Philip IV of Spain in Madrid.

REMBRANDT VAN RIJN
1606–1669
Dutch painter, draftsman, and etcher who worked mainly in Amsterdam.

**30-SECOND TEXT**
Francis Ames-Lewis

*Leonardo's portraits reflect the status of his sitters, but also crystallize the intimate engagement between sitter and beholder.*

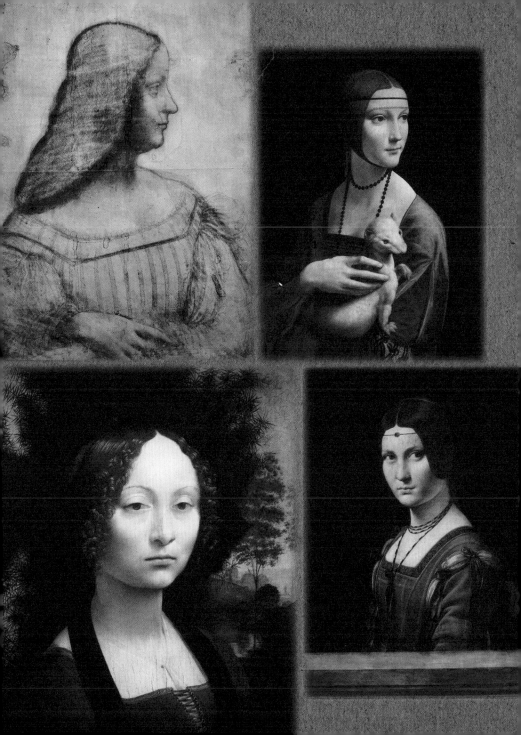

# DRAWING IN BLACK & RED CHALK

## the 30-second theory

## Natural black and red chalks

were quarried mainly in northern Italy. The material was cut into sticks, which could be sharpened to draw fine lines or used bluntly to generate areas of varying tone. Chalk is the ideal graphic medium for producing sfumato effects—delicate tonal gradients applied to model forms in three dimensions. While he was an assistant in Andrea del Verrocchio's workshop between the mid-1460s and mid-1470s, Leonardo must have known his master's black chalk drawings. Early on in his own drawing practice, however, he did not exploit chalk's potential to generate monochromatic tonal transitions. It was only when making studies in the mid-1490s, for the heads of apostles in the *Last Supper*, that Leonardo fully exploited the medium, becoming the earliest Western draftsman to use red chalk. The sharp, linear, red-chalk outline of Judas's profile combines with gently graded tonal modeling that brings out his neck musculature, exaggerated by emotional tension. Leonardo used the softer, more friable black chalk for his study of St. Philip to generate a more delicate atmosphere for this expressive face. Fluidly curving lines describe the strands of hair, and gentle rubbing of the chalk defines shadow in the eye sockets and below the jaw.

**RELATED RESEARCH**
See also
CHIAROSCURO & SFUMATO
page 22

THE LAST SUPPER
page 24

**3-SECOND SKETCH**
Leonardo found that black and red chalks were the ideal media for making detailed studies of smoothly modeled, emotionally expressive facial types.

**3-MINUTE MASTERPIECE**
In his *Libro dell'Arte* of ca. 1400, Cennino Cennini refers to black chalk, but this medium was little used before the late 15th century. Leonardo was the first to exploit red chalk's coloristic warmth, which was much valued by later draftsmen such as Correggio and Rubens. The range of Leonardo's chalk drawing practice inspired generations of draftsmen from the 16th century to the "academic" artists of the 19th century.

**3-SECOND BIOGRAPHIES**
CENNINO CENNINI
ca. 1370–ca. 1440
Florentine painter, and writer of the technical treatise *Il Libro dell'Arte* (The Craftsman's Handbook).

ANDREA DEL VERROCCHIO
ca. 1435–1488
Florentine painter, draftsman, and sculptor, in whose workshop Leonardo was apprenticed.

ANTONIO ALLEGRI CORREGGIO
ca. 1490–1534
North Italian painter and draftsman, active mainly in Parma.

**30-SECOND TEXT**
Francis Ames-Lewis

*Leonardo selected the harder lines of red chalk over softer, more tonal black to convey heightened emotion.*

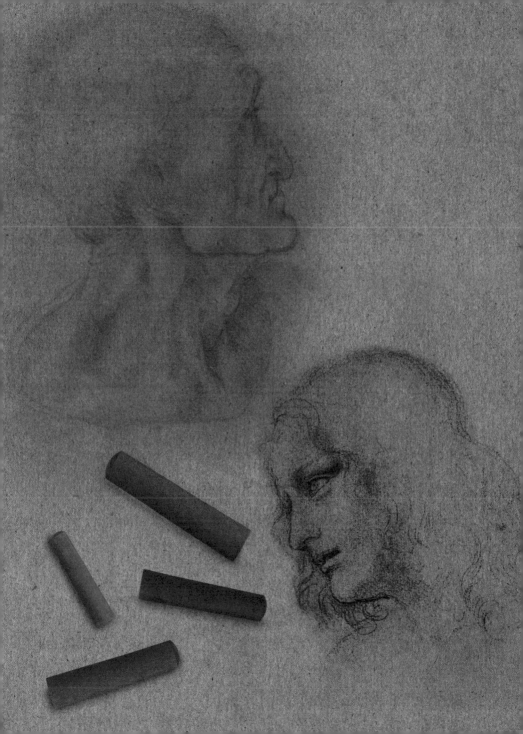

# THE DELUGE

## the 30-second theory

**Toward the end of his career,** from around 1515, Leonardo became preoccupied with a highly inventive vision of "the Deluge." He explored this cataclysmic evocation of nature's ability to wreak havoc through hurricanes, earthquakes, and other natural phenomena in both word and image. His lengthy written description, which evolves from advice on representing a battle, dwells upon the destructive power of the elements. A turbulent landslide destroys great trees. "Mountains collapse headlong into the depths of a valley." A river bursts its dam and the resulting waves demolish the valley cities. Flashes of lightning illuminate these catastrophic events. These and other violent episodes are reflected in an extraordinary series of large-scale, black chalk drawings. In one, the swirling movements of gale-force wind and torrential rain seem to cause a mountain to fragment and collapse onto an extensive and minutely depicted town. In others, dynamic, helical tornadoes (demonstrating Leonardo's constant fascination with spiraling movement) hurl themselves across the sheets, as though his imagination had torn itself free of his normally rational mind. Scarcely a single figure is to be seen—in the face of these natural disasters, man is insignificant, too small and too helpless to be worth recording.

**RELATED RESEARCH**
See also
WATER
page 128

FORCES OF NATURE
page 132

**30-SECOND TEXT**
Francis Ames-Lewis

**3-SECOND SKETCH**
A series of drawings of cataclysmic deluges that Leonardo made late in life indicate that his imagination was still vivid, but now perhaps disturbed.

**3-MINUTE MASTERPIECE**
What triggered these apocalyptic fantasies? Following the French invasions of the 1490s, Italy was unstable. Leonardo himself suggested that the wrath of God lay behind his images of destruction—he was perhaps disturbed by the Protestant reformers' threats to papal authority and the social order. He worked for two unproductive, insecure years in Rome when "the Deluge" fired his imagination, while his health was deteriorating.

*"The Deluge" is far more than Leonardo's artistic imagining of elemental fury: here, too, is a scientist's observation of how winds spiral out of storm clouds and spread as a destructive vortex ring once they reach the ground.*

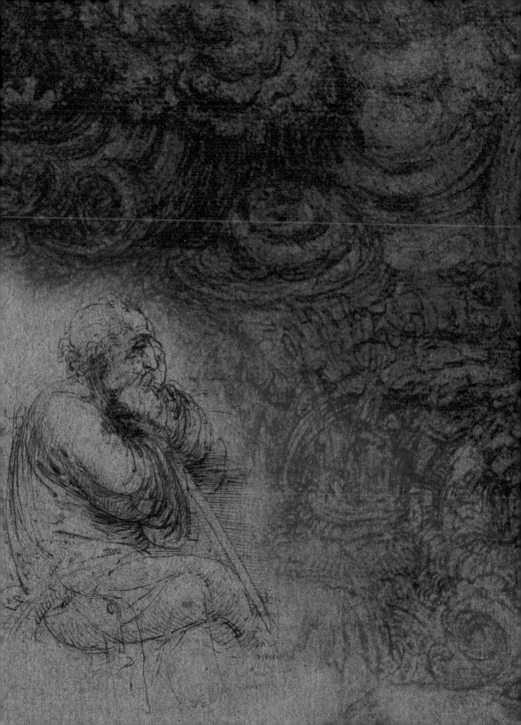

# GEOMETRY

**Archimedean solids** Semiregular polyhedra whose faces are regular polygons of two or more types that meet in a uniform pattern around each corner, named for Archimedes, who described them.

**atmospheric perspective** The visual effect of moisture and dust in the atmosphere on landscapes, particularly noticeable on distant objects, where colors seem to become bluer and tones more muted—the farther the object, the more its color matches that of the surrounding atmosphere. In painting, this gives an impression of depth and distance and is achieved by using true colors for foreground objects, gradually muting both colors and tones, as well as decreasing outline sharpness for the depiction of objects in the background. Also called aerial perspective.

**centroid** The center of mass of a two-dimensional geometric object of uniform density. Also known as geometric center.

**golden ratio** 1:1.61803—a ratio that is believed to be the equation for things perceived as being aesthetically pleasing. It can be found in ancient world designs such as the Pyramids and the Parthenon, and also commonly occurs in nature. Leonardo called it "the mean ratio" and is one of the geometrical ratios which are not expressible precisely in arithmetical terms or in numbers. The *Last Supper* is divided in golden ratio rectangles; and, most famously, the *Vitruvian Man* is proportioned according to this number. The golden ratio continues to influence the arts, from architecture and painting to industrial design.

**linear perspective** Until the early 15th century, painters didn't take into account the effects of perspective. In his 1435 treatise, *De Pictura*, the Florentine architect Leon Battista Alberti described in detail a mathematical system that gave a sense of depth to a flat surface by using optical phenomena, such as the apparent convergence of parallel lines and diminution in size of objects sited farther away from a fixed central viewpoint, that of a spectator looking at the painting. This system, called linear, optical, or mathematical perspective, was used until the late 19th century as the foundation of Western art.

**modulor** The French architect Le Corbusier (1887–1965), in the tradition of the *Vitruvian Man*, developed a scale of proportions that he called *Le Modulor*, based on a stylized 6-foot (1.83 m) man whose height is divided in golden ratio sections to provide "a harmonic measure to the human scale, universally applicable to architecture and mechanics."

**orthogonal lines** Parallel lines in a drawing that come together at the vanishing point. From a Greek expression meaning at right angle (to the front plane of a painting).

**perspectograph** An optical device invented by Leonardo to help give accurate linear perspective to a drawing. It consisted of a table with a stand that had a cutout used as a viewing window, and in front of it a clear sheet of glass on which the artist could sketch a scene with all elements in proper relation to each other. The drawing was then reproduced on canvas as an outline and the details painted in.

**Platonic solids** Regular polyhedra discovered by the Pythagoreans but later described by the Greek philosopher Plato, who speculated that these solids were the shapes of the fundamental components of the physical universe. Each face is the same regular polygon shape and each angle has the same value. There are only five Platonic solids—tetrahedron, cube, octahedron, dodecahedron, and icosahedron.

**vanishing point** According to the principles of linear perspective, first a line is drawn across the picture plane where the sky meets the ground—the horizon line—and a point is selected near the center of the horizon line—the vanishing point—where all parallel lines in the drawing will converge.

# LINEAR PERSPECTIVE

## the 30-second theory

### Leonardo identified three kinds

of perspective: of size (now called "linear perspective"), and of color and disappearance (now known, collectively, as "atmospheric perspective"). Together they describe the appearance of distant objects as smaller, less distinct, paler, and bluer. Linear perspective was developed by Brunelleschi, Alberti, and Piero della Francesca, and Leonardo perfected it. It is a geometric system for representing objects on a flat surface, the "picture plane," like tracing a distant scene on a window. Leonardo's perspectograph would help an artist to make a perspective drawing in that manner. All linear perspective theory is based on the fact that an object looks smaller as its distance from the viewer increases—what Leonardo called "proportional diminution with distance"—the object's size eventually shrinking to zero at what is referred to as the "vanishing point." This is well-illustrated, for example, in Leonardo's unfinished *Adoration*, which features a tiled floor, the lines ("orthogonals") of which converge to the vanishing point. Similarly, his *Annunciation* shows a perspective framework of incised lines on the wood panel on which it is painted, while in the *Last Supper* the vanishing point is placed at the level of Jesus's right eye, and his arms even lie along the orthogonals.

### 3-SECOND SKETCH

*"Perspective is the rein and rudder of painting."* This statement, and countless journal entries by Leonardo, show how important this concept was to him.

### 3-MINUTE MASTERPIECE

Leonardo's later studies of vision altered his earlier assumptions about linear perspective: that a light ray does not come straight into the eye but is refracted by the lens surface; or that a long wall parallel to the picture plane drawn in traditional perspective would be rectangular with sides parallel to the sides of the picture, but it is actually distorted and perhaps even curved. Other complications to perspective were binocular vision, optical illusions, and the earth's curvature.

**RELATED RESEARCH**
See also
THE LAST SUPPER
page 24

**3-SECOND BIOGRAPHIES**
FILIPPO BRUNELLESCHI
1377–1446
Leading architect and engineer of the Italian Renaissance who discovered perspective. Engineered the dome of the Florence cathedral.

LEON BATTISTA ALBERTI
1404–1472
Italian author, artist, architect, and poet.

PIERO DELLA FRANCESCA
1420–1492
Early Renaissance painter, mathematician, and geometer.

**30-SECOND TEXT**
Paul Calter

*Leonardo applied his understanding of geometry to his early paintings to achieve the illusion of space and distance in a two-dimensional plane.*

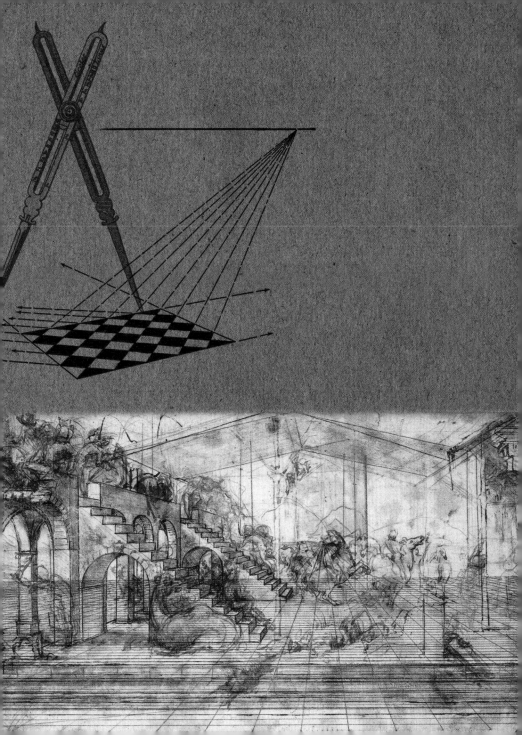

# GEOMETRIC OPTICS

## the 30-second theory

**3-SECOND SKETCH**
Leonardo's painting and
his interest in perspective
went hand in hand with his
study of optics and vision.

**3-MINUTE MASTERPIECE**
Leonardo created a model
of sight in which light
enters a chamber through
a pinhole and is projected
onto a screen; this
functioned as a camera
obscura (literally "dark
chamber"). Light from
an outside scene passes
through a small hole in one
side of the chamber and
forms an inverted image
on the opposite side.
Leonardo was fascinated
by it and described it in
his notebooks. There is
evidence that other artists
used the camera obscura.

Leonardo wrote: "the eye, which is said to be the window of the soul, is the primary way in which the sensory receptacle of the brain may ... contemplate the infinite works of nature." Dissecting eyes, and drawing a diagram of one, he rejected the medieval notion that the eye sees by sending out light rays instead of receiving them. In a bid to understand the mechanics of vision, he experimented with lenses, mirrors, apertures, and glass balls filled with water. One experiment demonstrated that small objects close to the eye are not visible. Another showed that the edges of a body cannot be seen clearly. A third, called the "moving needle illusion," revealed how a moving object close to the eye appears to move in the opposite direction. He also showed that objects underwater do not appear to be in their true place due to the refraction of light. Leonardo wrote extensively about shadows and advised artists to view their paintings reflected in a mirror. He made diagrams explaining the reflection of light rays from both flat and curved mirrors, invented an apparatus for making a curved mirror, and may even have contemplated making a "burning mirror" for military use.

**RELATED RESEARCH**
See also
LINEAR PERSPECTIVE
page 38

GEOMETRIC PROPORTIONS
page 42

**3-SECOND BIOGRAPHIES**
DAVID HOCKNEY
1937–
English painter, draftsman,
printmaker, stage designer,
and photographer. His book,
*Secret Knowledge*, published
with Charles Falco, discusses
the use of optical devices in
the history of art.

CHARLES M. FALCO
1948–
American experimental
physicist and expert on thin
film materials.

**30-SECOND TEXT**
Paul Calter

*As rigorously as he
studied anatomy to
achieve accuracy in his
drawings and paintings,
Leonardo strove to
transfer his scientific
observations of optical
phenomena to his art.*

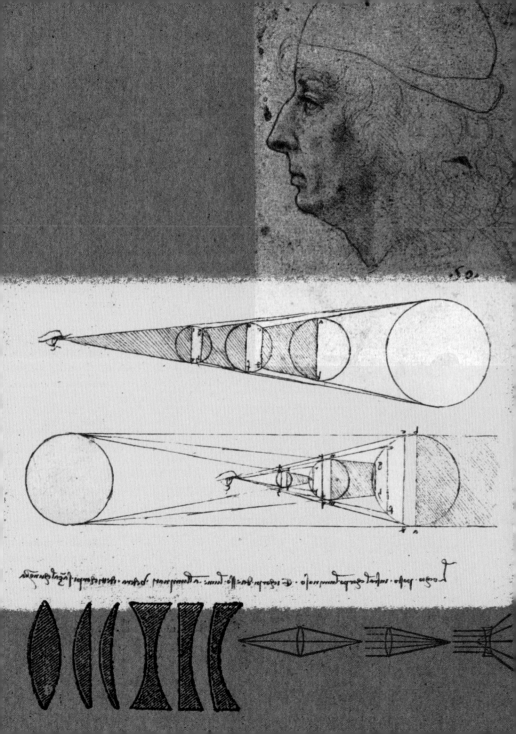

# GEOMETRIC PROPORTIONS

## the 30-second theory

### Leonardo's "Family tree of

proportions" charts the various mathematical proportions. It includes the "divine proportion," or golden ratio, which arises when a quantity is divided into two parts in such a way that the smaller part is to the larger part as the larger part is to the whole. Leonardo wrote that every part of the whole must be proportionate to the whole. This applies to people as well as animals and plants. He described the proportions of a man kneeling, standing, and seated, and he also observed that four fingers make one palm, four palms make one foot, six palms make a cubit, and four cubits make a man. Moreover, he calculated that an arm span equals a person's height, that the distance from the hairline to the bottom of the chin is one-tenth of a person's height, that from the bottom of the chin to the top of the head is one-eighth of a person's height, and so on for the rest of the body. Leonardo learned from the Roman architect Vitruvius that if a person lies flat with their arms raised to the level of their head, with their legs extended so that the space described by the legs makes an equilateral triangle, a circle drawn with the navel as its center would touch the fingers and toes.

**3-SECOND SKETCH**
Leonardo studied mathematical proportions, but as a painter he was mainly drawn to geometric proportions, particularly those of the human body.

**3-MINUTE MASTERPIECE**
From Vitruvius, Leonardo learned that a person not only fits into a circle but also that the distance from the soles of the feet of (unstretched) legs to the top of the head is the same as the span of the arms when held horizontally, describing a square. This is shown in Leonardo's iconic *Vitruvian Man*. This was taken as the link between the organic and geometric bases of beauty, and it also provided fuel for speculation about squaring the circle.

**RELATED RESEARCH**
See also
SQUARING THE CIRCLE
page 48

SOLID GEOMETRY
page 50

**3-SECOND BIOGRAPHY**
MARCUS VITRUVIUS POLLIO
ca. 80–70 BCE–ca. 15 BCE
Roman architect, engineer, and author of *De Architectura*.

**30-SECOND TEXT**
Paul Calter

*"Man as a measure of all things."* The version of **Vitruvian Man** Leonardo created in **1490 reveals his abiding interest in proportion, the study of which fused his artistic and scientific objectives.**

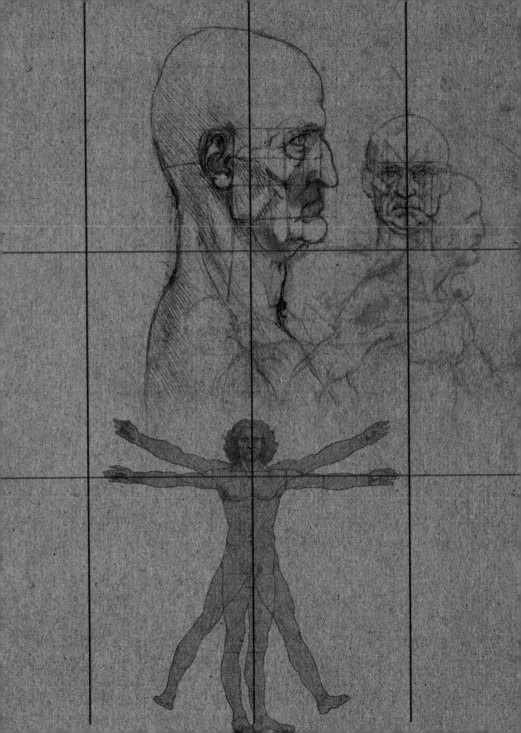

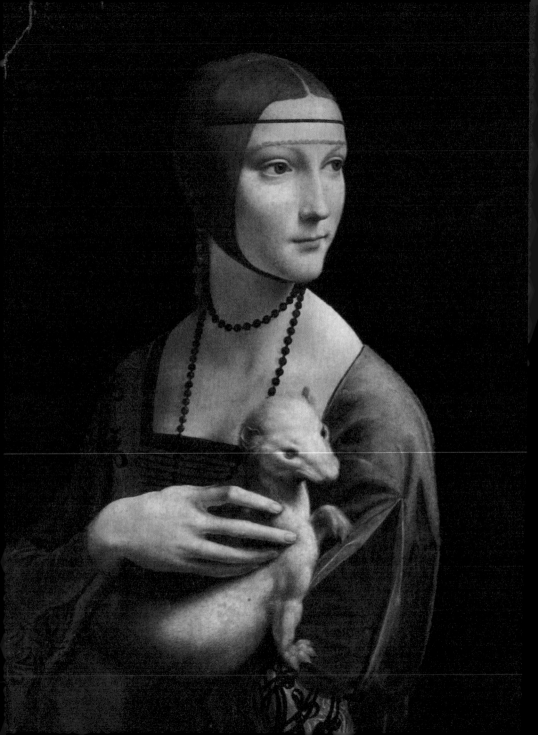

# LADY WITH AN ERMINE

Departing from the tradition of depicting high-ranked sitters in profile, the *Lady with an Ermine* occupies a central place in Leonardo's revolutionary conception of portraiture. The sitter shows an unprecedented sense of human communication. She is depicted in a three-quarter view, turning to her left as if reacting, with sparkling eyes and the beginnings of a smile, to someone or something next to her outside the picture frame. The animal she is holding, an ermine, looks attentively in the same direction and raises a paw, which reinforces her own movement and emotional response. The portrait was first praised by court poet Bernardo Bellincioni for the remarkable naturalism and sense of vitality of the sitter: "she appears to be listening." That effect is also achieved by the sophisticated system of light and color, seen also in the *Virgin of the Rocks* (see page 122). In creating areas of direct and reflected light of different intensities, the piece can be seen as a visual expression of some of Leonardo's theoretical precepts in the *Treatise on Painting*. The subtlety of the gradation from light to shade on the sitter's face, shoulders, and hand, from the directly illuminated to increasingly darker areas that receive oblique light, creates a remarkable sense of three dimensionality. Similarly, the fall of light on the ermine's soft coat produces a wonderful effect of tonal modulation and almost a dissolution of form in the darker passages, while the lighter areas cast secondary light onto the lower part of the sitter's hand. Given these stylistic qualities—and the identification of the sitter as Cecilia Gallerani, a mistress of Duke Ludovico Sforza—the portrait can be dated to ca. 1489–90. It bears, in addition, several levels of symbolic meaning that would have been particularly appreciated in the courtly context in which it was to be viewed. According to medieval bestiaries and Leonardo's own observations in his notebooks, the ermine stands for purity and moderation, which presumably evokes Cecilia's qualities. As the Greek for ermine is *galée*, the animal would also have been a pun on her family name. It may well also allude to Ludovico, awarded with the chivalric Order of the Ermine in 1488, and who is referred to by Bellincioni as the "white ermine." Highly regarded, Leonardo's painting was sent to Isabella d'Este, the Marchioness of Mantua, to be compared with works by Giovanni Bellini, and most likely influenced her subsequent commission for Leonardo of her portrait.

*Juliana Barone*

# PLANE GEOMETRY

## the 30-second theory

**3-SECOND SKETCH**
"Let no man who is not
a mathematician read
the elements of my
work." Leonardo's words
show how important
mathematics, especially
geometry, was to him.

**3-MINUTE MASTERPIECE**
Leonardo's studies in
geometry were applied
to his other areas of
interest. In the area of
military weapons, he
sketched the parabolic
path of projectiles, with
suitable allowance for
wind resistance, and in
anatomy he performed a
geometrical analysis of a
heart valve. In architecture,
he systematically
determined the symmetries
of polygonal churches,
showing that he had a
complete grasp of all
the possibilities.

Leonardo's notebooks reveal his intense study of Euclid's *Elements*, many theorems of which are illustrated in his notes. Some of his own contributions include a proof of Pythagoras' theorem (one of hundreds of proofs of this theorem). Farther, while most constructions of an angle of 15 degrees are fairly complicated, Leonardo showed how to do a "rusty compass" construction, that is, one with a fixed setting. He also devised a novel way to draw an ellipse using a triangular cutout. If the triangle is moved so that one vertex always lies on a vertical line and another vertex is always on the horizontal, the third vertex traces an ellipse. Leonardo was interested in the transformation of one body into another without adding or subtracting material, which formed the basis of his book *On Transformation*. Squaring of the circle is a perfect example of this. He also corrected Archimedes' erroneous version for finding the center of gravity of a trapezium (a four-sided figure with no parallel sides). Finally, his star octagon contains ratios of the parts that provide a modulor, that is, a series of proportions that ensures the repetition of similar shapes when used as a tool for architectural proportion.

**RELATED RESEARCH**
See also
SOLID GEOMETRY
page 50

ECCLESIASTICAL
ARCHITECTURE
page 80

**3-SECOND BIOGRAPHIES**
PYTHAGORAS
ca. 580–500 BCE
Greek geometer and
philosopher, one of the
greatest mathematicians
of his time.

EUCLID
ca. 300 BCE
Greek geometer,
astronomer, and physicist.
His *Elements* is the most
enduring mathematical
work ever written.

**30-SECOND TEXT**
Paul Calter

*Leonardo's layouts for polygonal churches shows his understanding of mirror symmetry and rotational symmetry.*

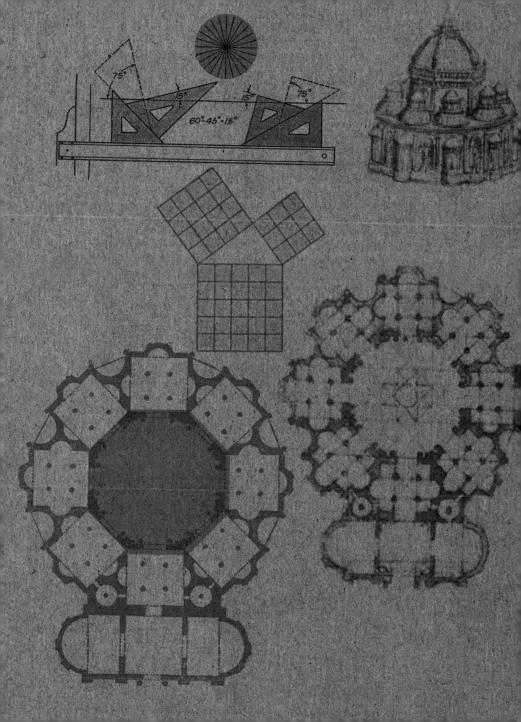

# SQUARING THE CIRCLE

## the 30-second theory

**3-SECOND SKETCH**
Squaring the circle—drawing a square with the same area as a given circle, using just a compass and straightedge—was a problem that engrossed Leonardo.

**3-MINUTE MASTERPIECE**
Leonardo was interested in constructing "geometric equations," to show the equivalence between plane figures with circular boundaries and those with straight boundaries. This was the idea behind his "Science of equiparation" and his "Book of equation." Within this is contained the larger idea that nature is always using the same amount of matter to produce an endless variety of forms.

**Leonardo's interest in squaring** the circle may be legitimately linked to his drawing of the *Vitruvian Man*, whose outstretched arms and legs fit into both a square and a circle. Leonardo took inspiration from Hippocrates' findings which stated that the area of a crescent or lune can be squared; that is, a square can be drawn that has exactly the same area as the crescent. If a crescent (a plane figure bounded by two circular arcs) can be squared, then why not a circle (which is bounded by just one circular arc)? Pages of Leonardo's manuscripts are covered with diagrams showing the areas of crescents, segments and sectors of circles, falcates (triangles with truncated ends), and rosettes. He devised a triangular technique for squaring a circle. He also used the method of inscribing a regular polygon within a circle, then subdividing the areas between the circle and a side of the polygon. He even started a book called *De Ludo Geometrico*, a collection of geometric diversions related to these figures. Leonardo was elated when he thought he had solved the famous problem, but his solution turned out to be incorrect, and it was proved much later to be impossible.

**RELATED RESEARCH**
See also
GEOMETRIC PROPORTIONS
page 42

**3-SECOND BIOGRAPHIES**
HIPPOCRATES
ca. 460 BCE–ca. 370 BCE
Greek mathematician and physicist, first to write a systematically organized geometry textbook.

MARCUS VITRUVIUS POLLIO
ca. 80–70 BCE–ca. 15 BCE
Roman architect, engineer, and author of *De Architectura*.

**30-SECOND TEXT**
Paul Calter

**Vitruvian Man, *based on the correlations of ideal human proportions with geometry, may have been influenced by Francesco di Giorgio, a contemporary of Leonardo, who used human proportions in his architectural treatise.***

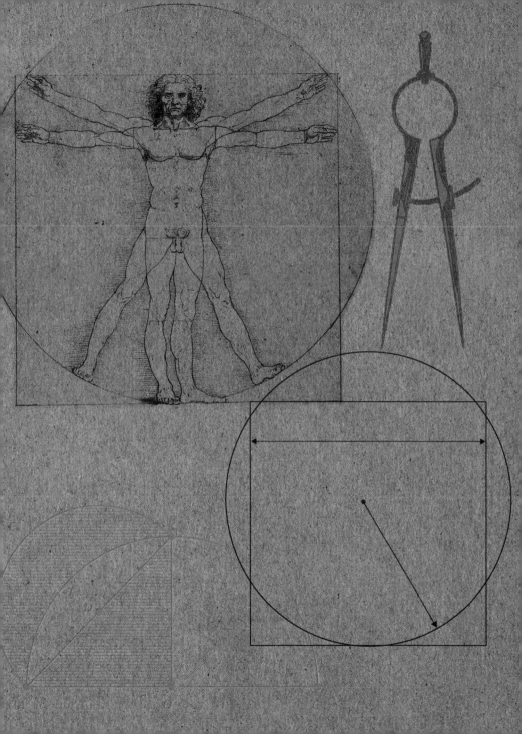

# SOLID GEOMETRY

## the 30-second theory

**In his drawings, Leonardo** showed how to transform four of the Platonic solids (the tetrahedron, octahedron, icosahedron, and dodecahedron) into the fifth, a cube of equal volume. He also demonstrated that all pyramids that have the same base and height have the same volume. Leonardo extended his method for finding the center of gravity of a trapezium to that of finding the center of gravity of a tetrahedron (a pyramid with a triangular base). He determined that the center of gravity is found one-quarter the way up from the base on a line joining the centroid of the base to the opposite vertex. He also devised a way to duplicate the cube, solving the so-called "Delian problem." Starting with a cube of a given side, the aim is to find, using only a compass and straightedge, the side of a cube having twice the volume. In his drawing of the solids for *Luca Pacioli's* book, *De Divina Proportione*, he depicts the Platonic solids mentioned above, some of the Archimedean solids (obtained by cutting off the corners of the Platonic solids), and some stellated or star-shape solids (in which adjacent faces are extended until they meet at a point).

**RELATED RESEARCH**
See also
GEOMETRIC PROPORTIONS
page 42

**3-SECOND SKETCH**
Leonardo's study of solid bodies was aided by his unique ability to visualize and manipulate geometrical shapes in a concrete way.

**3-MINUTE MASTERPIECE**
Leonardo provided 60 drawings of solids for *Pacioli's* book, *De Divina Proportione*. The same polyhedron is often shown both in solid and skeletal form, indicated by the labels *solidus* or *vacuus*. For the hollow figures he devised a brilliant new form of geometric illustration using "solid edges," which makes it easy to see which edges are in front and which are behind, as these can easily be confused in a simple line drawing.

**3-SECOND BIOGRAPHY**
LUCA PACIOLI
ca. 1445–1519
Renowned Italian mathematician, lecturer, teacher, and prolific author. Called the "Father of Accounting" for his invention of double-entry bookkeeping.

**30-SECOND TEXT**
Paul Calter

*Leonardo enlisted his artistic genius to draw complex mathematics. In his depiction of Archimidean and stellated solids, he provided a plane view and an empty view (illustrated here) in which the sides were removed to show the complete structure of the polyhedra.*

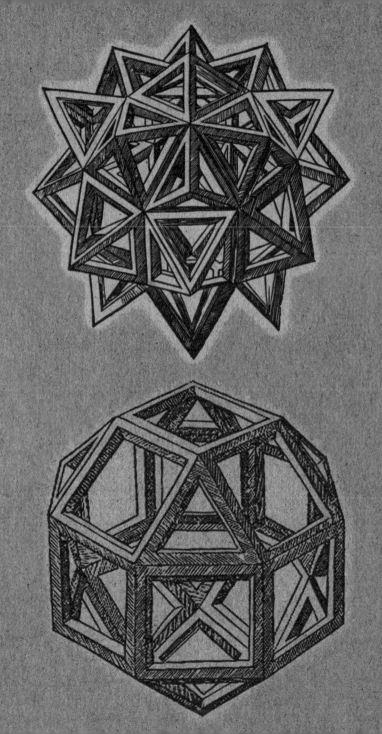

# KNOTS & ROSETTES

## the 30-second theory

**3-SECOND SKETCH**
Leonardo devised
decorations of knots and
rosettes, notably also in
the shape of tree branches,
following an elaborate
system of patterns of
interlaced designs similar
to those used in Islamic,
Celtic, and Byzantine art.

**3-MINUTE MASTERPIECE**
The Sala delle Asse in
the Castello Sforzesco in
Milan is decorated with
Leonardo's extensive
arboreal ornamentation
of intricate and elaborate
patterns. The painting of
the upper part of the walls,
extending to the whole
ceiling, is rich in decorative
elements also clearly
playing with mathematical
models of knot and
interlaced motifs.

Ornaments, such as knots and
rosettes, are common decorative motifs in
Medieval and Renaissance paintings and by no
means exclusive to the Western world. They
were used in Islamic decorations—arabesques—
and employed much more widely (they appear
regularly in Celtic and Byzantine designs as well
as in many other patterns). They are thought
to have meant different things in different
cultures. Designs using a single component, and
often a circular or spiraling pattern, may have
variously represented continuity and eternity
in different cultures. Leonardo's practice of
including decorations, such as knots and
rosettes, and interlaced designs in his paintings,
drawn with a purposely made compass,
must be seen in relation to his interest in
mathematics and geometry. One of Leonardo's
long-lived concerns was squaring the circle.
He devoted hundreds of pages in his notebooks
to this question, which he called *quadratura*.
In mathematical language, a knot is an
embedding of a circle in a three-dimensional
Euclidian space. It seems, however, that
Leonardo's investigations, including his
production of patterns such as knots and
rosettes, achieved no appreciable gain for
mathematics, although they did create a wealth
of complex and pleasing designs.

**RELATED RESEARCH**
See also
SQUARING THE CIRCLE
page 48

**3-SECOND BIOGRAPHY**
ALBRECHT DÜRER
1471–1528
German Renaissance artist
who produced a series of
woodcuts of knot patterns
(Six Knots) as direct copies
of engravings from the circle
of the Academy of Leonardo
da Vinci.

**30-SECOND TEXT**
Marina Wallace

*Applying characteristic
devotion to geometric
experiment, Leonardo
understood that the
underlying, repeating
patterns of the natural
world were informed
by mathematical
principles.*

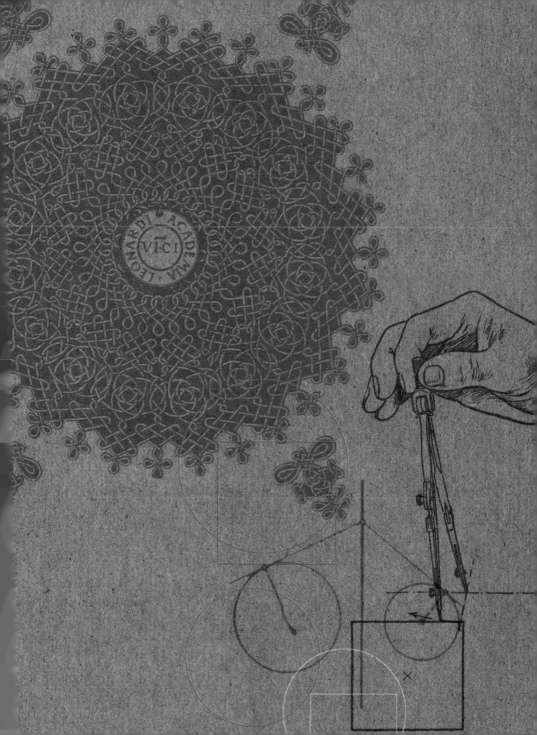

# MECHANICS

**aerofoil** A structure with curved surfaces that causes air to flow faster over the top than under the bottom. This creates a greater pressure below the aerofoil than above it, and the difference in pressure produces lift.

**Archimedes' screw** A water pump invented by Archimedes consisting of a screw inside an inclined cylinder whose base sits in water. When the screw is turned, water is lifted and flows to the top of the cylinder. A reverse Archimedes' screw, where water flowing through the cylinder forces the screw to turn, now powers a generator supplying electricity to Windsor Castle, in the UK.

**cartoon** From the Italian *cartone*, meaning a "large sheet of paper," a cartoon is a full-size drawing in preparation for a painting in fresco or on canvas. The cartoon was laid on the final surface and the drawing transferred by first pricking the lines of the composition with tiny holes and applying powdered charcoal to leave black dots that outlined the drawing; a process known as "pouncing."

**Codex Atlanticus** Kept at the Ambrosiana Library in Milan, the Codex Atlanticus is the largest collection of drawings by Leonardo. The work comprises 1,119 pages in 12 volumes, with 1,751 drawings and 100 pages written in Leonardo's right-to-left, mirror-image script. The name Atlanticus comes from its very large format, $25\frac{1}{2}$ inches (64.5 cm) by $17\frac{1}{4}$ inches (43.5 cm), known as Atlante. Written during his most creative years (1478–1519) the Codex captures the genius of Leonardo in its most eclectic form and demonstrates his contribution to science, engineering, and invention, as well as fine arts and literature. Painstakingly restored between 1962 and 1972 by the Basilian monks at the abbey of Santa Maria di Grottaferrata near Rome, the manuscript is now available for viewing online.

**escapement mechanism** In his Codex Madrid, Leonardo noted that: "It is custom to oppose the violent motion of the wheel of the clock by their counterweights with certain devices called escapements. They regulate the motion according to the required slowness and the length of the hour."

**Manuscript B** Kept at the Institut de France in Paris, Manuscript B was written in 1487 and 1489 and is the earliest bound notebook by Leonardo. It contains designs of mechanical inventions, such as his helicopter and other flying machines, as well as a submarine and several war machines. It is written in Leonardo's right-to-left, mirror-image script. Although originally bound together, it is now on separate sheets.

**Newton's third law** In his 1687 *"Philosophiae Naturalis Principia Mathematica,"* Sir Isaac Newton presented his three laws of motion. The third law states that "to every action there is always an equal and opposite reaction: or the forces of two bodies on each other are always equal and are directed in opposite directions."

**peg and cage gears** A peg or toothed gear consists of a disk with pegs radiating from its circumference; a cage or lantern gear is made up of two disks on the same axle connected by long cylindrical pegs. These gears are easy to make and are an efficient way of transmitting rotary motion.

**rack and pinion** A mechanism consisting of a single gear (the pinion) engaging with a sliding rack. It is an efficient way to convert rotary motion into a back-and-forth motion and, even today, has many applications, such as the windshield wipers of a car.

**ratchet and pawl** A mechanism that allows a wheel to turn in only one direction. It comprises a wheel with specially shaped teeth (the ratchet) and a bar on a pivot (the pawl) fixed above the ratchet wheel. The pawl slides over the teeth of the ratchet in one direction, but stops the motion of the teeth if the wheel attempts to turn in the other direction.

# HELICOPTER

## the 30-second theory

**Although usually described as a** helicopter, Leonardo's elegant flying machine design from 1483 (appearing in one of the folios of Manuscript B) is really an aerial version of an Archimedes' screw. The design was for a helical linen wing stretched on a spiral of iron wire supported by a wooden structure. It was almost certainly never built. The number "8" appears on the diagram, suggesting the device was intended to be 8 braccia across. This measure, based on the human arm, is usually taken as around 26 inches (66 cm), making the screw 17 feet (5.3 m) across, though it could be anything between 11 feet 10 inches (3.6 m) and 18 feet 8 inches (5.7 m). The device was powered by its four human passengers pushing on rods that turned the central shaft. As the wing rotated, it would push down the air, producing an equivalent upthrust, which we now know would be thanks to Newton's third law. Although the machine could never actually fly—a screw is not an ideal way to turn rotary motion into thrust in air, and four men could only provide a fraction of the energy required to lift themselves this way—its beautiful, iconic form is an inspiration for rotary-powered flight and remains one of Leonardo's best-recognized designs.

**3-SECOND SKETCH**
Though not a direct ancestor of the modern helicopter, Leonardo's aerial screw provides a very clear pointer to the value of rotating aerofoils.

**3-MINUTE MASTERPIECE**
Leonardo commented: "If this instrument made with a screw is well made—that is to say, made of linen of which the pores are stopped up with starch—and is turned quickly, the screw will make a spiral in the air and it will rise high." In fact, his design for a spit, powered by a turbine in the chimney, was a more realistic air-driven concept, using a rack and pinion to rotate the meat.

**RELATED RESEARCH**
See also
PARACHUTE
page 62

**3-SECOND BIOGRAPHIES**
GE HONG
238–343 CE
Chinese writer who describes something a little like a helicopter.

MIKHAIL LOMONOSOV
1711–1765
Russian scientist who demonstrated a prototype rotor device.

PAUL CORNU
1881–1944
French inventor who made the first helicopter free flight.

**30-SECOND TEXT**
Brian Clegg

*Although his aerial screw never got off the ground, in his lifetime Leonardo successfully advanced the understanding of disciplines such as civil engineering, optics, and hydrodynamics.*

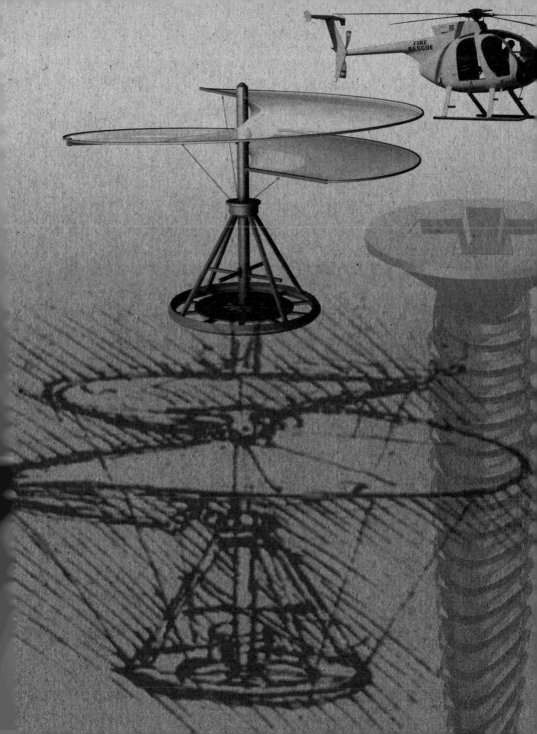

# WORM GEARS

## the 30-second theory

Although gears don't have the exotic appeal of a programmable self-propelled cart, they were central to the success of the Industrial Revolution and Leonardo was a pioneer in their use, sprinkling them liberally throughout his manuscripts. Sometimes he employed simple peg and cage gears, but he was at his most sophisticated in his use of worm drives. These gears, which Leonardo called "endless screws," use a long screw rod engaging on a circular gear to switch rotation up to 90 degrees. They had been around for a while, but Leonardo points out that the circular gear needs a pawl and ratchet to stop it from rolling back, in case the single gear tooth engaged with the worm breaks: "Endless screws that engage only one of the teeth of the working wheel could cause great damage and destruction if the tooth breaks." Leonardo's design uses a worm that tapers in the middle so that it engages with several of the gear teeth at the same time, doing away with the need for the ratchet. This mechanism was reinvented by English clockmaker Henry Hindley and is now known as a Hindley gear, despite Leonardo's 300-year precedent. Leonardo also made a pair of long worm gears useful in devising the powerful double screw jack, used for lifting heavy objects.

**3-SECOND SKETCH**
By devising a worm gear with a waist, Leonardo increased safety and efficiency in a design that would not be matched for nearly 300 years.

**3-MINUTE MASTERPIECE**
Leonardo not only concerned himself with the design of worm gears, but thought through how to manufacture them. In Manuscript B, he shows an elegant device to cut a long screw or worm gear, which includes several forward-thinking features, such as double guiding screw rods for stability, and interchangeable guide wheels to alter the pitch of the screw thread along the gear.

**RELATED RESEARCH**
See also
HYDRAULIC SAW
page 66

SELF-PROPELLED CART
page 72

TANK
page 102

**3-SECOND BIOGRAPHIES**
ARCHIMEDES
ca. 287–212 BCE
Devised both an odometer and an orrery that would have needed gears.

HERO OF ALEXANDRIA
ca. 10–70 CE
First recorded description of the use of a gear, though they were employed for at least three centuries before.

**30-SECOND TEXT**
Brian Clegg

*The myriad technical drawings in Leonardo's notebooks attest to his predominant interest in mechanical devices, increasingly at the expense of his art.*

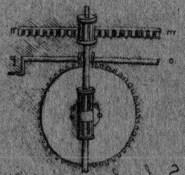

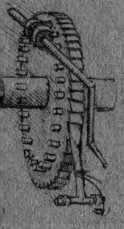

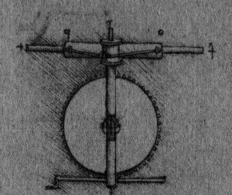

*Leonardo da Vinci — mechanical studies (mirror-script notes)*

# PARACHUTE

## the 30-second theory

Leonardo produced a range of flying machine designs, often with complex mechanisms to duplicate the movements of a bird's wing. But it was his simplest design from the Codex Atlanticus that was the most practical. The parachute is a large piece of linen cloth held open by a pyramid-shape frame of wooden poles. Although Leonardo's sketch suggests that the parachute is around the height of a man, the notation describes the poles of the frame as being 23 feet (7 m) long. Though not the first such design—there is a conical fabric-and-wood parachute for escaping burning buildings in an Italian manuscript from around ten years earlier—Leonardo's is the first workable design. It is interesting that Leonardo describes the parachute as a "tent"—anyone who has attempted to put up a tent in a high wind could sympathize with the idea that this experience was the inspiration for the design. The most worrying aspect of Leonardo's parachute is that there is no obvious harness. The passenger is hanging from raised arms, grimly holding onto cords attached to the frame. This may be because the original is a simple sketch instead of one of Leonardo's more complex designs—it is probable that he would have added some form of harness to a more detailed consideration of the requirements.

**3-SECOND SKETCH**
Leonardo's parachute, designed so that a man "can jump from any great height whatsoever without injury," demonstrates a simple but practical design that would actually work.

**3-MINUTE MASTERPIECE**
The design has been tested in practice by sky diver Adrian Nicholas, who jumped from a hot air balloon at almost 9,840 feet (3,000 m) high in 2000. He discovered that the design was effective and gave a smoother descent than a modern parachute, but the weight from the frame (a common feature of all early parachutes) and heavy fabric, around 187 pounds (85 kg), would endanger the passenger on landing, and Nicholas switched to a modern parachute for the last part of the drop.

**RELATED RESEARCH**
See also
HELICOPTER
page 58

**3-SECOND BIOGRAPHIES**
FAUSTO VERANZIO
1551–1617
Designed an enhanced parachute from Leonardo's illustration and is reputed to have tested it.

LOUIS-SÉBASTIEN LENORMAND
1757–1837
The French inventor who coined the term "parachute."

JEAN-PIERRE BLANCHARD
1753–1809
Invented the frameless, folded silk parachute.

**30-SECOND TEXT**
Brian Clegg

*Twenty-first century tests using modern, lightweight materials confirm that Leonardo's original design could parachute a man to the ground— albeit at the whim of the wind.*

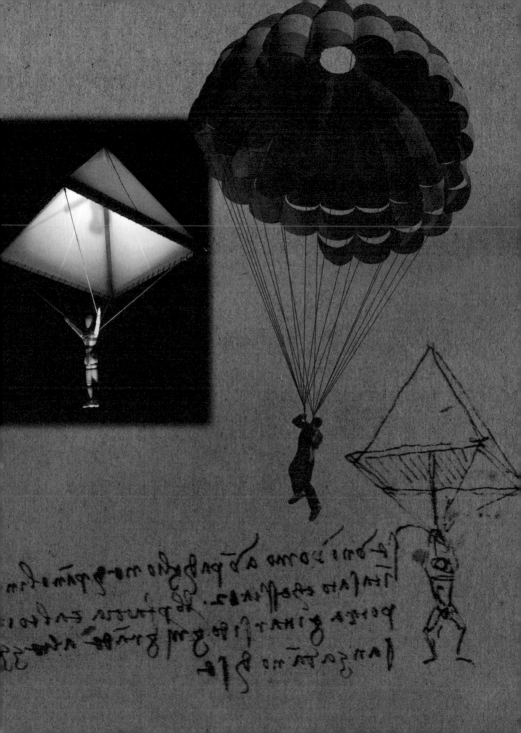

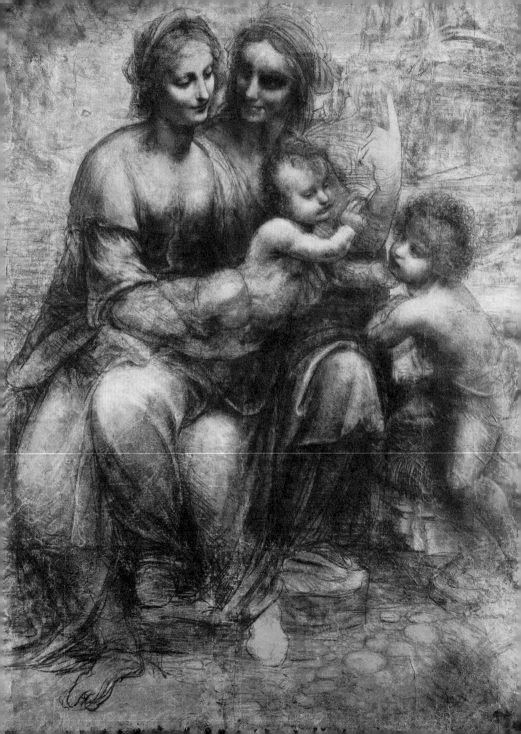

# THE VIRGIN AND CHILD WITH ST. ANNE AND THE YOUNG ST. JOHN THE BAPTIST

One of Leonardo's most beautiful works, this large-scale drawing (cartoon), now kept in London's National Gallery, is also one of the finest examples of his chiaroscuro technique and ideas on motion and emotion. By sketching the main outlines of the figures on brown paper and gradually modeling them with black and white chalks, he simultaneously defined the forms and adjusted the poses via pictorial effects. His skillful blending of shadow and light renders particularly well both the dynamism of the twisting movements and the emotional interaction of the figures through facial expression. The subject occupied Leonardo for about 17 years (ca. 1501–18). It is not known if he started work on the St. Anne on his own initiative, as a demonstration of his artistic skills upon his arrival in Florence, or on a commission from the Servite friars with whom he was lodging. It may be that the advent of King Louis XII's patronage in 1507 prompted revisions of the composition. Most likely dating from around that time, the National Gallery cartoon offers a different composition from that of an earlier cartoon, now lost, but described by Pietro da Novellara in 1501. What distinguishes one cartoon from the other is the presence (or absence) of the lamb, and the orientation of the composition to the left (or right). In the National Gallery cartoon, the lamb is replaced by John, who is being blessed by Christ, while Mary, seated on Anne's lap, turns toward the right to adjust herself to the twisting movement of her son. Although never used by Leonardo himself for a painting, the cartoon was available in Milan at the beginning of the 16th century and was carefully copied by workshop assistants and followers. By contrast, the St. Anne painting by Leonardo that is now in the Louvre bears considerable similarities with the 1501 cartoon, though this painting seems to have been started a few years later and reworked until the last years of his life. Leonardo's compositional variations and revisions are typical of his open, creative process, in which his "brainstorming" method of jotting ideas on paper provided him with a range of possibilities, which he continuously explored in a nonlinear sequence, leading to "finished" compositions intrinsically dependent on one another. The National Gallery cartoon emerges both as a crucial moment of consolidation of Leonardo's thoughts and as one of the most magnificent examples of the Renaissance "well-finished" cartoon (*ben finito cartone*).

*Juliana Barone*

# HYDRAULIC SAW

## the 30-second theory

**3-SECOND SKETCH**
The hydraulic saw is a simple but clever design to use the rotary motion of the water wheel to drive both the saw and the carriage for the wood to be cut.

**3-MINUTE MASTERPIECE**
The annotations for the saw design provide an interesting difference from much of Leonardo's work. The inscription "*Vuole essere più lungo tutto*" (roughly: "everything wants to be longer") is not in Leonardo's usual mirror writing, but written conventionally from left to right. This suggests that the diagram was not intended for his own use, but to communicate to someone else—or even that the criticism was not added by Leonardo.

The hydraulic saw designs in the Codex Atlanticus are not among Leonardo's most original ideas, but they do give an effective insight into the way he drew inspiration for later and more unique work. The saw is powered by a flow of water through an angled wooden channel, which drives a water wheel with boxlike paddles. The mechanism converts the rotating motion of the paddle wheel into a reciprocal action, driving a saw blade up and down. It also generates a constant pull on a carriage, through a series of pulleys. This wheeled container holds the piece of wood to be sawn, which is slowly drawn through the device. Saws of this kind were just coming into use at about the time the drawing was made in the late 1470s, and though Leonardo's design may have some original features, it was probably based on existing saws or drawings. This seems to be reflected in the way that the image has been sketched—with a more labored style than Leonardo's usual work. The same page also had some copied images of the turnbuckles popular with architects of the time, suggesting that this was the young Leonardo gathering ideas and information in preparation for his own inventions.

**RELATED RESEARCH**
See also
SELF-PROPELLED CART
page 72

TANK
page 102

**3-SECOND BIOGRAPHIES**
AUSONIUS
ca. 310–395 CE
Roman poet referred to the sound of sawmill (probably hydraulic) cutting marble.

VILLARD DE HONNECOURT
fl. 1250
Produced an early sketch of a water-driven sawmill.

ALBERT KAUFMANN
fl. 1946
Replaced the needle in his wife's sewing machine to invent the jigsaw, a reciprocal saw like Leonardo's.

**30-SECOND TEXT**
Brian Clegg

*A master of mechanical principles, Leonardo's saw designs enriched the centuries-old tradition of harnessing the power of water.*

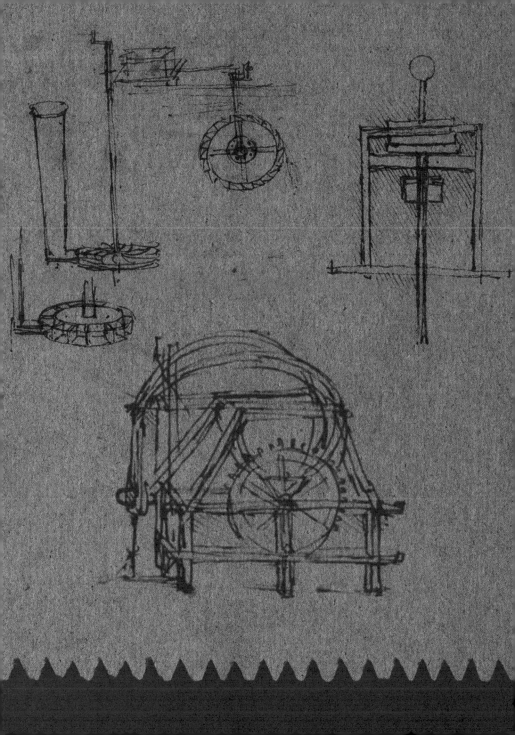

# VIOLA ORGANISTA

## the 30-second theory

### Dating from the 1490s,

Leonardo's viola organista is a fascinating portable keyboard instrument (predating the Geigenwerk of 1575). His drawings for it in red chalk, contained in the Codex Atlanticus, show many detailed parts for this original and ingenious device. A rod that descends from the instrument's body is strapped to the player's leg. As he walks along, this keeps a flywheel inside the mechanism in motion, which drives a continuously moving loop of horsehair, replacing the bow of a stringed instrument. As the player presses a key on the three-octave top-mounted keyboard, it pushes the string to be played against the "bow." The result is the reverse of playing a conventional instrument, where the bow is pressed against the string to produce a sound. The tone and volume could be modified by the amount of pressure applied to the key. The idea of producing a sound from using a constantly moving "bow" was already well-established with the hurdy-gurdy, which is more like a conventional stringed instrument in shape, but has a rotating wheel to do the bowing. However, Leonardo transformed this concept both in the way it was powered and through the much-expanded true keyboard, to produce a new class of instrument that had never before been conceived.

**3-SECOND SKETCH**
By combining a bowed string instrument with a keyboard, Leonardo produced a versatile portable string orchestra.

**3-MINUTE MASTERPIECE**
Part of the appeal of Leonardo's design is the way that he delves into small details in his intricate drawings. For the viola organista, he does not only cover the detailed mechanical construction of the instrument but also provides more than one illustration of the way the instrument would be strapped onto the player's body, leaving both hands free to access the keyboard.

**RELATED RESEARCH**
See also
WORM GEARS
page 60

**3-SECOND BIOGRAPHIES**
HANS HAIDEN
ca. 1540–1613
Constructed the first bowed, true keyboard instrument like Leonardo's, called the Geigenwerk.

AKIO OBUCHI
fl. 1969–
A Japanese keyboard maker who has produced a number of Geigenwerk-type instruments since the 1990s.

**30-SECOND TEXT**
Brian Clegg

*Pragmatic and inventive, Leonardo considered the needs of the player as well as the instrument's sound.*

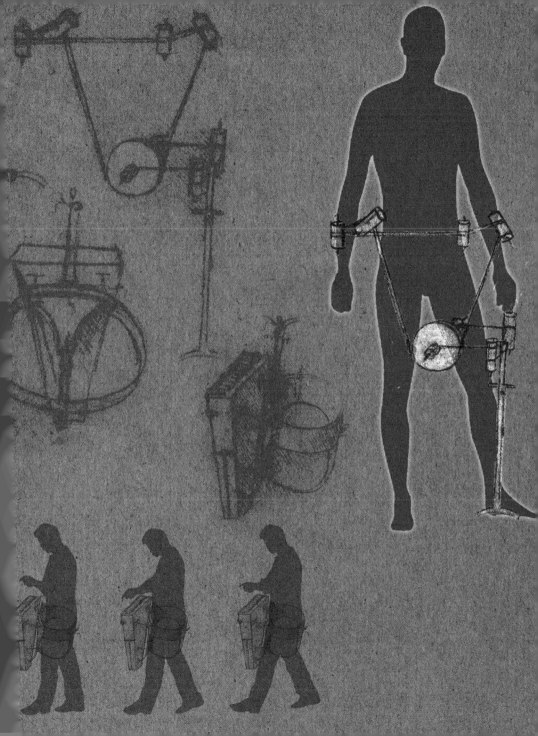

# DIVING SUIT

## the 30-second theory

### Leonardo's inventions are

remarkable because of the range of ideas he had. This becomes clear when looking at something as practical and yet original as his diving suit. We don't usually think of clothing as mechanical, but this is really an underwater breathing device, where Leonardo thought through the requirements of a diver and designed an apparatus to satisfy them. Air is sent down from the surface through bamboo tubes, joined with pig leather, feeding a leather face mask fitted with glass lenses. At the top is a large bell-shape float, which keeps the air inlets above the water. The pipes may seem clumsy, but we must keep in mind that, before the invention of the aqualung, diving did involve a helmet linked by tubes to the surface. In a variant of the design, Leonardo took a step toward scuba gear, incorporating a leather wineskin to act as an air reservoir. Based at the time in Venice, which was under threat from the Turkish fleet, Leonardo designed his suits with the intention of enabling defenders of the city to attack invading ships by cutting holes in their hulls under the waterline. He commented: "Knowing the evil in men's hearts, they will learn how to kill men on the seabed."

**RELATED RESEARCH**
See also
NAVAL WARFARE
page 108

WATER
page 128

**3-SECOND SKETCH**
Leonardo devised both a diving helmet and a kind of scuba gear intending for them to be used in Venice as defense against naval attack.

**3-MINUTE MASTERPIECE**
Some of Leonardo's designs are impractical despite seeming plausible, such as his bird-wing flying machines, or the helicopter. But the diving suit is not only practical (a version was tested by diver Jacquie Cozens), but incorporated forward-thinking ideas, such as using steel rings on the air tubes to prevent them from being crushed by pressure, and, in a second design, providing a urine bottle to enable the diver to stay underwater for long periods of time.

**3-SECOND BIOGRAPHIES**
ARISTOTLE
384–322 BCE
Mentioned the use of a "cauldron" forced down into the water as a kind of diving bell.

JOHN LETHBRIDGE
1675–1759
Constructed what is often considered the first underwater diving apparatus.

JACQUES-YVES COUSTEAU
1910–1997
Popularized diving and coinvented the "aqualung" self-contained underwater breathing apparatus.

**30-SECOND TEXT**
Brian Clegg

*Of Leonardo's many inventions connected with water, few were as forward-thinking as his underwater breathing apparatus.*

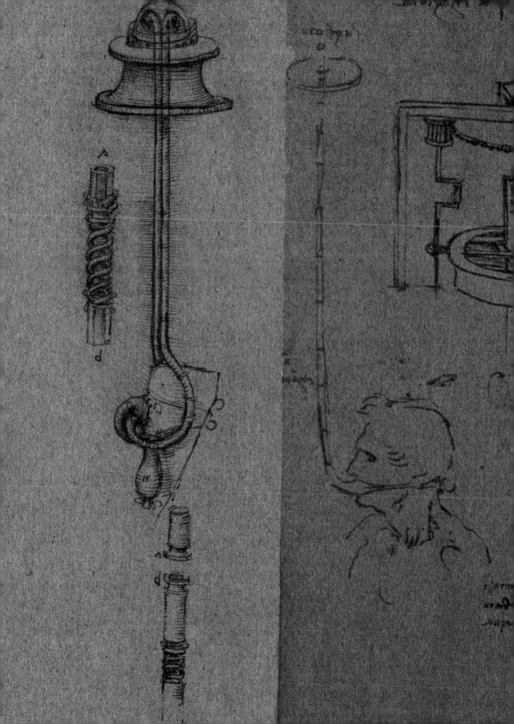

# SELF-PROPELLED CART

## the 30-second theory

### Once described as an

"automobile," Leonardo's design for a self-propelled cart, found in the Codex Atlanticus from around 1478, resembles a massive timepiece. Two large coiled springs provide the motive power—the added complication of the second spring recognizes the need to maximize the force to overcome the limitations of the relatively crude gearing. The two springs are wound in opposite directions and fitted into wooden drums, with their output linked by interlocking gears. Force is transmitted by a series of cams, driving a cage-shape gear. The strongest similarity to a clock is the way in which the power is transferred to the driving wheels by an escapement mechanism. This provides a more controlled release of power, one notch at a time, smoothing out the spring's natural tendency to weaken as it unwinds. The three-wheel cart, probably intended to be around 5 feet (1.5 m) across, is steered by a single front wheel that could be crudely programmed (though only to steer to the right) with a series of wooden blocks placed between the gears, making it operable without a human driver onboard. This turn is also evidenced by what appears to be a brake that can be operated remotely, holding the wound springs under tension until the brake is released by pulling on a cord or cable.

**RELATED RESEARCH**
See also
SPECTACULAR SIEGE
MACHINES
page 100

TANK
page 102

**3-SECOND SKETCH**
There is no evidence the cart was built, but the idea of a self-propelled and self-steering vehicle was hugely imaginative.

**3-MINUTE MASTERPIECE**
Although this was once considered to be a design for a car, the fact that the cart is designed to be operated remotely suggests that it may well have had a more dramatic application. Leonardo was known to have been involved in stage design that featured moving elements. It is quite possible the cart was intended as a mechanism for a stage effect.

**3-SECOND BIOGRAPHIES**
HERO OF ALEXANDRIA
ca. 10–70 CE
Described a self-propelled cart driven by weights and ropes, and mechanisms for theater.

FRANCESCO
DI GIORGIO MARTINI
ca. 1439–1502
Architect and painter who designed the "automobile," a human-powered vehicle.

FERDINAND VERBIEST
1623–1688
Flemish missionary to China credited with producing a model steam-powered car.

**30-SECOND TEXT**
Brian Clegg

*Leonardo's sketches for a cart with brakes and self-steering capabilities are, for many, a blueprint for the world's first robot.*

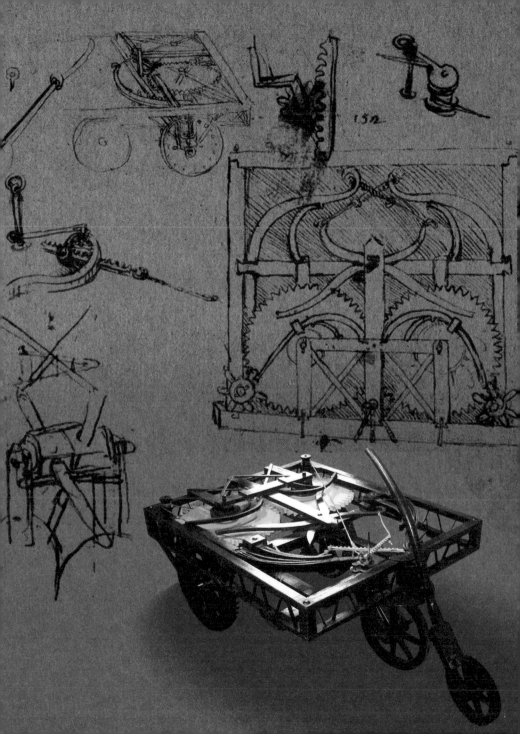

# CIVIL ENGINEERING

**canalization** The conversion of a river or natural stream into a continuous artificial waterway suitable for inland navigation or irrigation and drainage, by means of weirs, barrages, and locks.

**cartography** The art, technique, or practice of compiling or drawing maps or charts. In order to win Cesare Borgia's patronage, Leonardo drew a town plan of Imola in 1502, pacing the distances personally for precise measurement and accurate layout of the streets and fields around the town. Although maps had been in existence since the Babylonians, they were very rare and inaccurate, and Leonardo's new scientific approach to making maps impressed Cesare Borgia so much he employed him as his chief military engineer and architect.

**Codex Arundel** It was a Codex of miscellaneous sheets from various dates, now unbound. The unbound collection of papers (most dating to 1508) written in Italian by Leonardo in his characteristic right-to-left mirror writing. The Codex Arundel includes diagrams, drawings, and brief texts, covering a broad range of topics in science and art, as well as personal notes. The Codex was acquired by the art collector and politician Thomas Howard (1585–1646), second Earl of Arundel, presented to the Royal Society by his son Henry Howard, and eventually purchased by the British Museum. The Codex Arundel is now kept at the British Library and can be viewed online as a digitized manuscript.

**lost-wax process** A method for casting bronze dating back thousands of years to ancient Egypt and still the best method for capturing intricate detail in metal objects. First an artwork is created, usually in wax. A two-piece wet-clay mold is then made and liquid wax poured into it to make a hollow copy of the original. This is dipped in liquid clay several times to build the final mold, which will be fired in an oven to harden. The wax melts and flows out of the mold. Molten bronze is poured in the mold, which will be carefully hammered out when cool to reveal an exact copy of the original artwork.

**pontoons** In his Codex Atlanticus, Leonardo describes a floating bridge consisting of a wooden walkway set on six flat-bottom boats (pontoons), pivoting on a vertical shaft set on the riverbank to allow for the bridge to open and close. A winch is used to close the bridge and park it in a cutout in the riverbank, while the flow of the current opens it by moving the pontoons downstream.

**quadripartite symmetry** A design that divides a square or a rectangle through its two primary axes into four equal and symmetrical quadrants.

**the Pantheon** From the Greek *pan* (meaning "all") and *theon* ("gods"), the Pantheon is a building dedicated to all Roman gods and was built by Emperor Hadrian on the site of Marcus Agrippa's original temple, where, according to Roman legend, Romulus, the founder of Rome, was seized at his death by an eagle and taken off into the skies with the gods. Considered the Romans' most important architectural achievement, it boasted the largest dome in the world until the 15th century, and it is still the largest unreinforced concrete dome ever built. Its harmonious appearance, due to a precisely calibrated symmetry (its diameter is exactly equal to its interior height) inspired Michelangelo to describe it as "the work of angels, not men."

*tiburio* In Medieval architecture, the crossing tower at the intersection of the nave, chancel, and transept of a church.

**topography** From Greek *topos* (meaning "a place") and *grapho* ("to write"), topography is a graphic representation of the surface features of a place, town, or region on a map, indicating their relative positions and elevations in order to create a three-dimensional picture.

# THE IDEAL CITY

## the 30-second theory

### Leonardo, as well as other

engineers and architects, devised ambitious plans and elevations for the ideal city. Around 1485, he was at work on an ideal plan for Milan that included an extensive canal system and the elevation of living quarters and gardens high above ground. It seems that he proposed this design of an ideal city to Ludovico Sforza—in response to the outbreak of the plague in Milan in 1485—as one of his attempts to gain long-term employment as an engineer for the court. Although he obtained this position in 1489, plans for the canalization and elevation of the city remained only on paper. Nevertheless, these plans were compared with those of the great civil engineers of Milan, including Filarete, Bramante, and esteemed guests, such as Francesco di Giorgio in 1490. By 1517, Leonardo's fame as an artist and engineer earned him a position at the French royal court of Francis I, for whom Leonardo produced initial plans and elevations for two royal palaces surrounded by rings of canals, as the center of an ideal city in Romorantin. Francis I told Benvenuto Cellini that Leonardo was a "man of some knowledge of Latin and Greek literature," a qualification Francis probably considered for the architect of a project that would identify Romorantin with the grandeur of ancient Rome.

**3-SECOND SKETCH**
A principal necessity of Renaissance civil engineering was that the greater good of a city involved the Humanist integration of civil projects with the city plan.

**3-MINUTE MASTERPIECE**
In early January 1517, Leonardo returned to his Milanese experiences when he was asked to develop plans for an ideal palace and canalization of Romorantin—a new Rome—which was to be a central hub for canals throughout France. Leonardo brought to this project his treatise on "sculpture, painting, and architecture" (now lost) and hundreds of architecture and canal studies, as well as the measurements of local streets by Francesco Melzi.

**RELATED RESEARCH**
See also
CARTOGRAPHY & CANALS
page 82

**3-SECOND BIOGRAPHY**
FRANCESCO FILARETE
(ANTONIO AVERLINO)
1400–1469
As an influence on Leonardo, Filarete's ideal plan for Sforzinda around 1457 remained a model for the Sforzas, who were fashioning Milan as the Athens of Italy.

**30-SECOND TEXT**
Matthew Landrus

*Designed to be pleasant as well as sanitary, the ideal city relied on a tiered system of upper walkways and lower roads, separating dwellers from a network of canals that provided transportation and removed waste, preventing the spread of contagious diseases that ravaged Europe.*

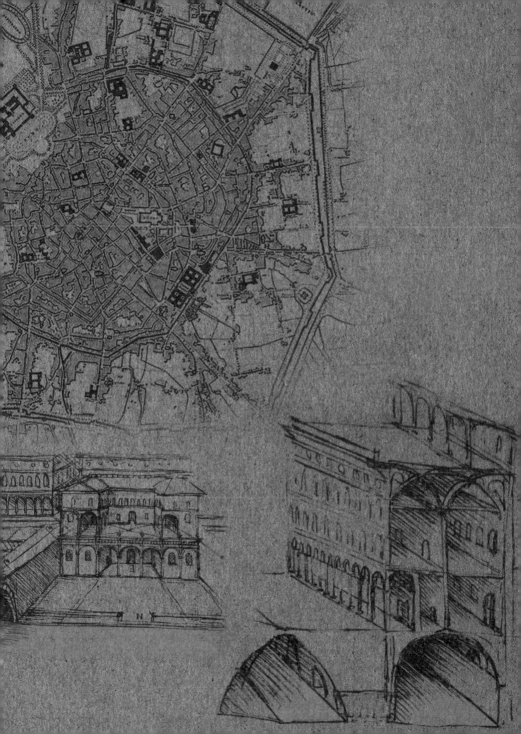

# ECCLESIASTICAL ARCHITECTURE

## the 30-second theory

**RELATED RESEARCH**
See also
GEOMETRIC PROPORTIONS
page 42

**3-SECOND SKETCH**
With the classical standards of Vitruvius and Leon Battista Alberti, Leonardo developed exceptional examples of Renaissance church structures and designs based on mathematically rigorous proportions.

**3-MINUTE MASTERPIECE**
Leonardo's church designs possibly influenced Donato Bramante. For example, Bramante's famous sculptured wall motif, wherein buildings appear to be sculpted blocks, recalls the form and niches of earlier church designs by Leonardo. In another of Leonardo's examples, the church's width is equal to its height and the precision quadripartite symmetry is interrupted only by the porch, reminiscent of the Pantheon (126 CE), the only building that was regarded as perfect during the Renaissance.

Following urban planning studies in his early notebooks (Manuscript B, the Codex Trivulzianus, and the Codex Atlanticus), Leonardo produced a range of designs for churches during 1487–90, most of which were on a central plan. This period of work coincided with his two competition entries for the construction of Milan Cathedral's *tiburio*, the covering for its dome. His innovative designs used triangular interlocking brick spines over a steep, acute arch dome. Most of the thrust, or lateral stress, was directed down onto triangular bases and compound piers. He employed Bernardo Maggi da Abbiate to build two wooden models for the competition, and possibly copied his original drawings for the carpenter by pricking (punching with tiny holes) them for transfer to other pages. Although he withdrew his model long before the commission was granted to two Milanese architects, Leonardo was nonetheless in the company of the exceptional architects Donato Bramante and Francesco di Giorgio in the *tiburio* competition and in other civil engineering projects. The central-plan churches in Leonardo's Manuscript B probably influenced Bramante's approaches to similar structures, even if basic plans for these central system churches on polygonal and circular-base geometry were initially developed in early 15th-century Tuscany.

**3-SECOND BIOGRAPHY**
DONATO BRAMANTE
1444–1514
A close friend of Leonardo's at the Sforza Court, who had been court architect since 1476. He designed several churches in the city, including the sacristy of Santa Maria delle Grazie, home of Leonardo's *Last Supper* wall painting in the monastery's refectory.

**30-SECOND TEXT**
Matthew Landrus

*The octagon was used repeatedly by Leonardo. In his domed church designs, which recall those of Brunelleschi and others, it is his geometric expression of grandeur allied with spatial practicality.*

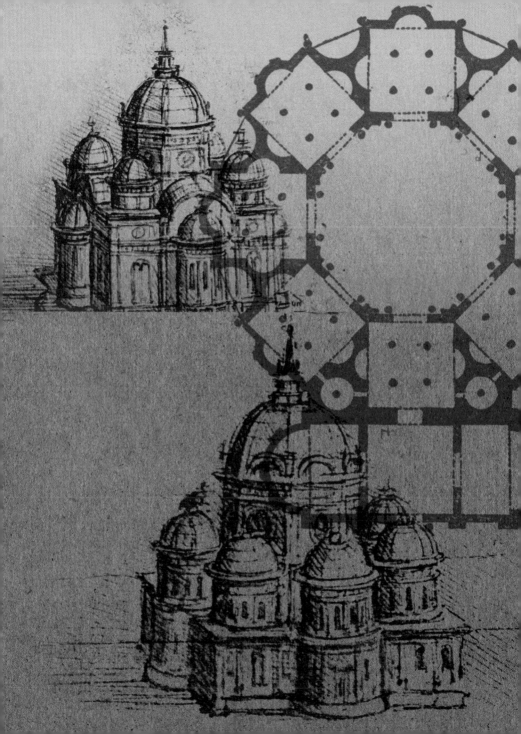

# CARTOGRAPHY & CANALS

## the 30-second theory

**3-SECOND SKETCH**
As an important source of income for Leonardo and his patrons, water resources were utilized and managed with impressive technological efficiency in Renaissance Italy.

**3-MINUTE MASTERPIECE**
During 1513–16, Leonardo produced a topographical map for Pope Leo X, with canals he proposed for draining the Pontine Marshes, southeast of Rome. To accompany his proposal, he had hundreds of technical illustrations from 35 years of study, enough for a formal treatise on the subject. In the Codex Atlanticus, for example, he illustrated the construction of a lock in 1485, to which he added details on the doors around 1493.

**Throughout his professional** career, Leonardo studied and designed canals and developed associated cartographic, topographic, and geographic studies. Almost every year of his notebook entries addresses the nature of water, its kinetic energy, and its regulation. He was also a respected hydraulic engineer: his most famous canal project was commissioned by the Florentine Republic in 1503 to divert water from the Arno River in Florence away from its natural course toward eastern Pisa. Were it to be excavated in northern Florence, an early stage of this abandoned canal could be his only surviving civil engineering project. For the Republic, he produced detailed maps, studies of canal locks, and designs for a giant canal- and lagoon-digging crane, which is illustrated in the Codex Atlanticus. In the previous year, Cesare Borgia appointed Leonardo "beloved court architect and engineer general," with the task of making precision maps and hydraulic studies of cities in Emilia, Tuscany, Umbria, and the Marches, including his famous map of Imola. Before his return to Milan in 1508, he confirmed with the French governor his earlier rights to the Naviglio Grande, a canal in southwest Milan, "to set up my mechanical devices and things which will most greatly please Our Most Christian King."

**RELATED RESEARCH**
See also
WATER
page 128

**3-SECOND BIOGRAPHY**
CESARE BORGIA
1475–1507
The illegitimate son of Pope Alexander VI. From 1498 until his death in battle in 1507, he was the ruthless Captain-General of the papal armies.

**30-SECOND TEXT**
Matthew Landrus

*Leonardo's hydraulic engineering projects included dams, canals, and ambitious drainage plans for which he additionally designed excavating equipment. He also undertook surveys to produce maps of such accuracy they gave his patron, Cesare Borgia, real military advantage*

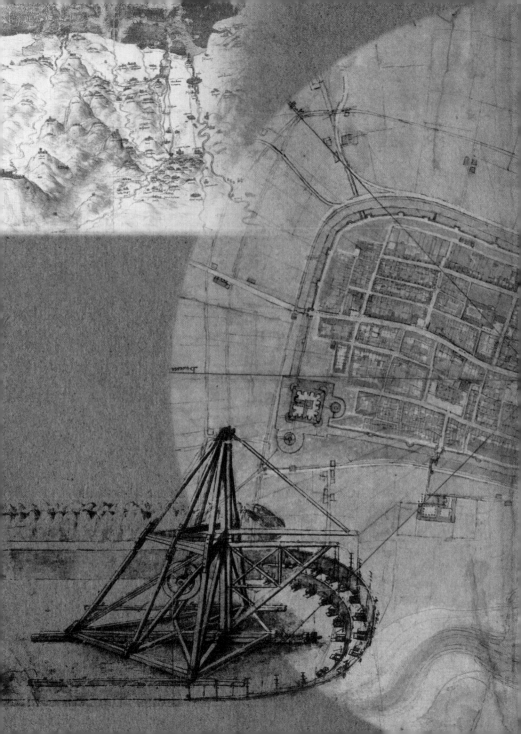

# BRIDGES

## the 30-second theory

### 3-SECOND SKETCH
Leonardo's elegant bridges stand up to modern engineering scrutiny and were breathtaking in the originality of their scale and design features.

### 3-MINUTE MASTERPIECE
One problem with wooden bridges that used curved beams is that the beam can split under load. In the Codex Atlanticus, Leonardo shows how a series of jagged notches in a beam can prevent splintering. It was more than 300 years before this innovation was used in practice in 19th-century Swiss bridges, such as the 1839 bridge in Signau, crossing the Emme River, that can take a load of 44 tons (40 tonnes).

In bridges, Leonardo found a civil engineering challenge equal to his skills. His designs included a revolving bridge for military use that was counterweighted to swing across a stream or moat, and an innovative double-decker bridge. But perhaps the greatest project Leonardo considered was a bridge to span the Golden Horn from Constantinople to Pera, a structure 1,150 feet (350 m) long. This beautiful, flowing design, sketched out in Manuscript L, appears to be a proposal for the Sultan of the Ottoman Empire. It is known that agents of Sultan Bayezid II traveled to Rome in 1502 in search of engineers to design a permanent bridge to replace the pontoons that were used across the Golden Horn at that time. Michelangelo is thought to be one of those approached, but according to a letter from Leonardo to the Sultan found in the archives of Topkapi Sarayi in Istanbul, it seems Leonardo also made a bid to design the bridge along with a windmill, a pumping device, and a drawbridge, most of which are also sketched in Manuscript L. A smaller footbridge by Vebjørn Sand, based on Leonardo's design, was constructed in 2001 outside Oslo in Norway, and Sand has established the Leonardo Bridge Project to inspire future structures around the world.

**RELATED RESEARCH**
See also
THE IDEAL CITY
page 78

CARTOGRAPHY & CANALS
page 82

**3-SECOND BIOGRAPHIES**
BAYEZID II
1447–1512
Sultan of the Ottoman Empire.

MICHELANGELO DI LODOVICO BUONARROTI SIMONI
1475–1564
Painter, sculptor, architect, and engineer. Young challenger to Leonardo's supremacy.

VEBJØRN SAND
1966–
Norwegian artist, founder of the Leonardo Bridge Project.

**30-SECOND TEXT**
Brian Clegg

*In the pay of the Duke of Milan, Leonardo devised many schemes, including emergency bridges using trees lashed together to get troops over rivers with an element of suprise.*

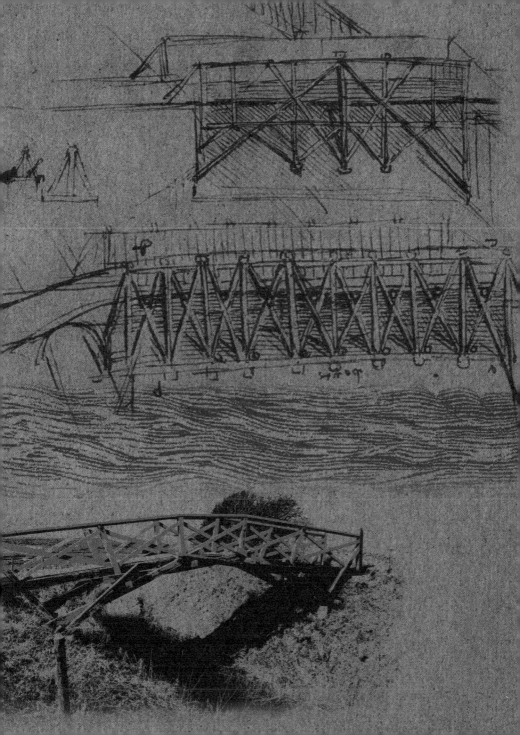

# CASTING THE SFORZA HORSE

## the 30-second theory

**Leonardo's first significant** commission for the Sforza court was for an equestrian monument in honor of Ludovico Sforza's father, Francesco. As was the court standard for Leonardo, no task was too ambitious for Lord Sforza, for whom he proposed a bronze horse nearly 33 feet (10 m) tall, weighing 85.1 tons (76.2 tonnes), to be completed by 1490. Precision drawings for the primary phase of the project reveal that Leonardo intended to cast the entire shell of the "colossus" instead of sections. He produced several presentation drawings of the molds, supportive structures, and machines necessary for a massive lost-wax casting process. As a structural engineer, he calculated the need for a hoist with ten stationary and nine moving pulleys, along with a special casting pit that included reverberatory furnaces (with separate smelting and burning chambers). A large clay model and tesserae—the terracotta mold sections—were produced by 1493, though by 1494 the bronze set aside for the project was sent to Ferrara for the casting of cannon. The Sforza Horse project that was first considered by Duke Galeazzo Maria Sforza in 1473, and reconsidered by his brother in 1483, was formally discontinued with the invasion of the French in 1499.

---

**3-SECOND SKETCH**
Although hired to sculpt the Sforza Horse, Leonardo developed plans for an equestrian "colossus" unlike any sculpture since antiquity, which also required his skills as a structural engineer.

**3-MINUTE MASTERPIECE**
In 1493, Leonardo installed at the Cathedral of Milan an effigy of the Sforza Horse, complete with horse, rider, and pedestal. Rising nearly 40 feet (12 m), this painted plaster, canvas, and wood panel construction honored the wedding for Habsburg emperor Maximilian I and Ludovico Sforza's niece, Bianca Maria Sforza. Court poets praised this ephemeral masterpiece. Court mathematician Luca Pacioli recalled in 1496 Leonardo's earlier plans to engineer a means for casting the massive bronze structure.

---

**RELATED RESEARCH**
See also
THE BATTLE OF ANGHIARI
page 88

**3-SECOND BIOGRAPHY**
FRANCESCO SFORZA
1401–1466
Illegitimate son of Muzio Sforza, and a famous military commander, community leader legislator, duke, and founder of the Sforza dynasty in Milan in 1450.

**30-SECOND TEXT**
Matthew Landrus

*Leonardo's numerous anatomical studies of horses drawn from life formed part of his bid to build the biggest equestrian statue the Renaissance world had ever seen. Only his drawings survive: his full-scale clay model was destroyed and the bronze monument itself never materialized.*

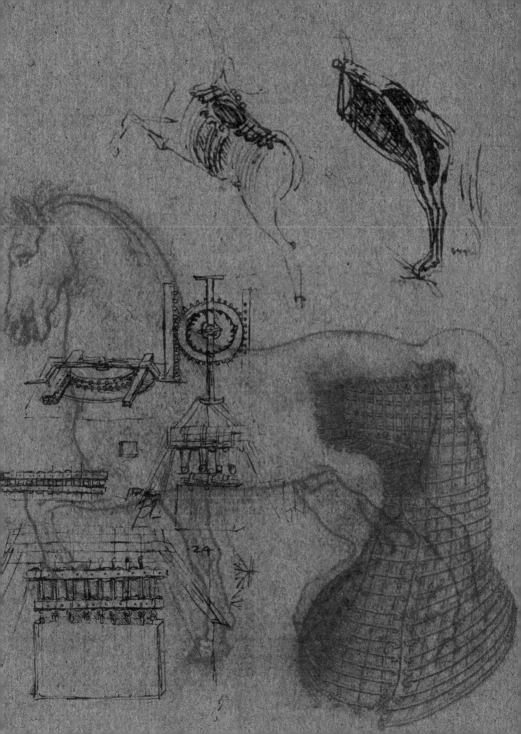

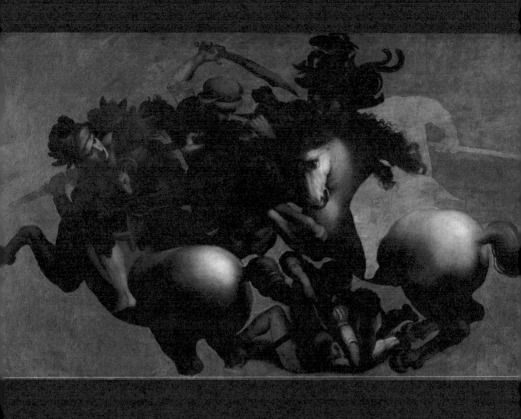

# THE BATTLE OF ANGHIARI

**Leonardo considered war the** "most beastly madness." He was commissioned to paint this subject on the wall of the Great Council Hall of Florence's civic palace (*Palazzo della Signoria*), in 1503. It was to represent the historic victory of the Florentines over the Milanese in 1440 at Anghiari. The work was to be part of a comprehensive decorative program aimed at celebrating the newly proclaimed Florentine Republic, and which included a painting by Michelangelo (*Battle of Cascina*). Leonardo's *Battle of Anghiari* is now lost. It is possible that what he actually painted was limited to the central scene, of which the Uffizi panel is one of the earliest copies and probably reflects the unfinished state of the painting. Centered on the frightful combat of men and horses, the crux of the fight is conceived as the moment when a bellowing soldier is about to deliver a furious blow to cut off the hand of his opponent, who is trying to get hold of the Milanese standard. Their rearing horses engage with no less violence, making use of their hooves and teeth. Leonardo's main biographer, Giorgio Vasari, gives a powerful description of the savagery of the combat, the terror, the anger, and the vindictiveness of both men and horses, in a fight in which the animalization of men is expressed in the parallels between human and animal physiognomies and expressions. Leonardo's painting was informed by his studies of motion, "physical motion" (*moto corporale*) and "mental motion" (*moto mentale*); and by his scientific investigations in comparative anatomy and physiognomy. Described as "the school of the world," the *Battle of Anghiari* did indeed influence artists ranging from Raphael and Michelangelo to Rubens. It was also copied by anonymous artists who reproduced the central group, or excised specific motifs (especially heads), in order to use them as components in paintings of other subjects. However, left unfinished and showing signs of early deterioration, probably due to Leonardo's experimental technique, the mural was covered up when the Hall was redecorated in the 1560s. Attempts at locating Leonardo's lost Battle continue; and a recent estimate of the size of the group he painted on the wall has been based on a cartoon of the *Head of the Shouting Soldier* in the Ashmolean Museum, in Oxford, England.

*Juliana Barone*

# THE STAGE

## the 30-second theory

**3-SECOND SKETCH**
Leonardo's stage designs were not simple sets, but instead inspired mechanical devices that brought the stage to life.

**3-MINUTE MASTERPIECE**
Leonardo's most impressive automaton was designed for a festival in Lyons in honor of the newly crowned King Francis I. As part of this festival, arranged by the merchants of Florence, a mechanical lion took the stage. It first walked forward, moving its head and tail, and then split open its chest to reveal lilies. In this astute political device, the lion of Florence was submitting by producing the symbol of the French king.

**In Leonardo's day, it was common** for architects, artists, and engineers to design for the stage. Leonardo was first exposed to this when apprenticed to Verrocchio, whose workshop produced sets and scenery for sacred plays performed for Galeazzo Maria Sforza in Florence. Leonardo's own stage work seems inspired by this experience, though he would focus on designing sets and automata for the secular theater. In his early work for the Sforza court, Leonardo stuck to the classically inspired perspective settings, though embellished with typical innovation—his design for the *Feast of Paradise* featured actors playing the planets, rotating in a large hemisphere. Later work was more naturalistic, moving away from classical strictures. This is best demonstrated by his remarkable design for a production of *Orpheus*, staged in Milan after the French invasion. Codices Arundel and Atlanticus show sketches of a series of mountains, one of which splits in two as the figures of Pluto and his minions rise from the underworld. Leonardo wrote: "When Pluto's Paradise opens, let there be devils with bellowing infernal voices." In typical Leonardo style, this effect involved a clever mechanism of counter-weights (which may have been stagehands), propelling the two halves of the mountain open and raising the infernal crew from the depths.

**RELATED RESEARCH**
See also
VIOLA ORGANISTA
page 68

SELF-PROPELLED CART
page 72

**3-SECOND BIOGRAPHIES**
AL-JAZARI
1136–1206
Kurdish scholar who devised mechanical automata.

ANDREA DEL VERROCCHIO
ca. 1435–1488
Italian sculptor and painter to whom Leonardo was apprenticed.

FRANCIS I
1494–1547
French monarch and Leonardo's final patron.

**30-SECOND TEXT**
Brian Clegg

*With his armor-clad automaton, Leonardo gave the world its first humanoid robot. Using a system of pulleys and cables, its head, jaw, and arms were capable of movement.*

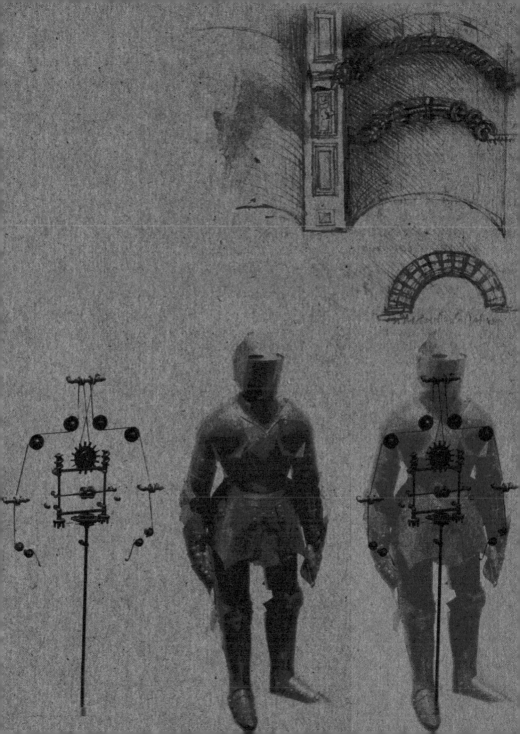

# INDUSTRIAL MACHINES

## the 30-second theory

**RELATED RESEARCH**
See also
WORM GEARS
page 60

HYDRAULIC SAW
page 66

**It is easy to focus on the more** glamorous aspects of Leonardo's work, although his machine designs probably improved the working lives of many. We tend to think of automation of industrial processes as a Victorian concept, but Leonardo's design for a file cutter shows the same idea in action. The operator turns a crank to raise a weight. After this the machine operates autonomously, using a worm gear to smoothly work along a metal bar that will turn into a file, while repeatedly raising and dropping a hammer that cuts the grooves in the surface of the file. Leonardo later used a similar approach in a complex six-way device for stamping designs in gold foil. Often, Leonardo's designs take a single physical movement by the operator and convert it into multiple actions, as in moving the file and the hammer. This was typified by his many designs for textile machines, including automated shearing devices, bobbin winders, and spinning machines. One particularly interesting device was for grinding convex mirrors. Leonardo dedicated considerable effort to researching the optical properties of these parabolic mirrors, using coded notes, probably to keep his ideas from Giovanni degli Specchi, a competitor working alongside him.

**3-SECOND SKETCH**
Leonardo's best industrial machines are designed to translate simple movement by the operator into a complex set of actions to automate a process.

**3-MINUTE MASTERPIECE**
Some of Leonardo's most modern-looking drawings are found in a study of the basics of industrial machines. His Codex Atlanticus shows a hoist that translates the backward and forward motion of a handle into the smooth rotation of large wheels to raise or lower a weight. Alongside the simple drawing is an exploded view to make it clear how the mechanism works. More than a single design, this seems to be an exercise in understanding reciprocal motion.

**3-SECOND BIOGRAPHIES**
FILIPPO BRUNELLESCHI
1377–1446
Designer of the dome of the Florence cathedral, finished by Leonardo's master, Verrocchio, with Leonardo's help, which required early industrial machinery to make the construction possible.

GIULIANO DE' MEDICI
1479–1516
Brother of Pope Leo X who employed Leonardo and his competitor Giovanni.

**30-SECOND TEXT**
Brian Clegg

*Leonardo's myriad industrial designs combine practicality with elegance. For him as an inventor, the noblest pleasure was the joy of understanding.*

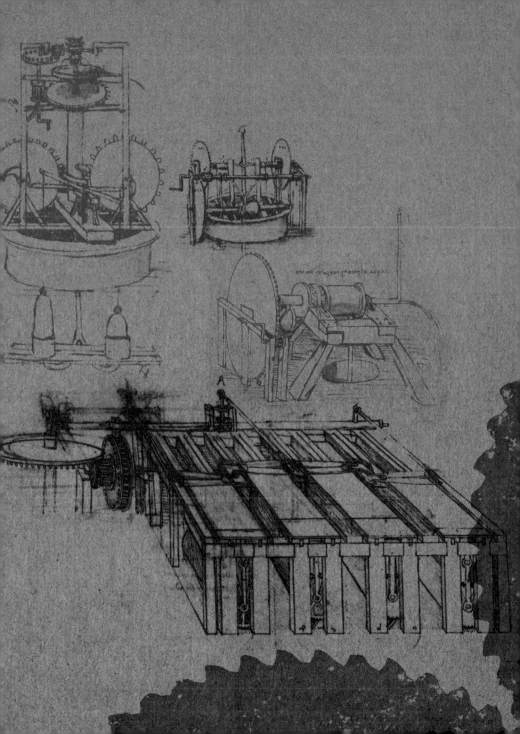

# MILITARY ENGINEERING

# MILITARY ENGINEERING
## GLOSSARY

*circumfolgore* From Italian, meaning "circular thunderbolt," the *circumfolgore* is described on the first page of the Codex Atlanticus. It is a formidable naval warfare machine, comprising a rotating platform mounted on the upper deck of a ship, on which were mounted 16 bombards (medium-size cannons). In a naval battle, the guns were constantly reloaded while the platform was rotated to point and fire the bombards, one at a time, at the enemy ship. This allowed for a very fast firing rate and greater precision compared to the usual layout of cannons pointing straight out of the sides of a ship.

**hydrodynamics** Literally "water movement," hydrodynamics is the branch of physics that examines the forces acted upon or exerted by liquids. Leonardo's manuscripts reveal that water and its movements were one of his most frequently studied and recorded topics. Among his drawings, there are many that are studies of the motion of water, in particular the forms taken by fast-flowing water on striking different surfaces and the effect of water in eroding rocks.

**merlon battlement** A battlement is a defensive wall on top of a castle, tower, or fortified house with openings (crenelles) and solid projections (merlons). Soldiers standing on the protected walkway behind the battlement could fire weapons and throw projectiles through the crenelles while hiding behind the merlons.

**mimetic conception** The knowledge that Leonardo gained from his studies and observation of nature was a source of inspiration for his inventions, many of which were based on the understanding, and imitation, of the principles of nature. His sketchbooks are filled with inventions that are closely linked to designs found in the natural world.

**polygonal fortress** Duke Cesare Borgia employed Leonardo as a military architect in 1502–3 to oversee all construction on his domain and design a fortress that would be strong enough to resist an important innovation in warfare of the period: firearms and increasingly powerful cannons. The Codex Atlanticus shows drawings of a polygonal fortress that, even now, could be considered very modern in its design, with inclined walls and sharp angles to deflect and absorb the impact of cannon balls. All the towers of the fortress are circular and, instead of the traditional battlement, have rounded tops that also help deflect the impact of cannon fire. Small openings make it possible for those defending the fortress to return fire with minimum risk. Leonardo's approach is backed by a thorough understanding of the strength of material and careful scientific observations.

**scoppietti** Sixteenth-century cannons used to take a long time to load and Leonardo solved the problem by inventing a multi-barreled gun consisting of a number of small-bore tubes (*scoppietti*) mounted on a single gun carriage on wheels. His 33-barrel organ (so called because the rows of barrels resemble those of an organ) had 3 racks of 11 guns arranged in a fan shape. While the first rack was fired, the second was cooling and the third was reloaded, and this made it possible to achieve a firing rate of up to ten shots per minute.

**springald** A Medieval catapult, similar in principle to a crossbow, capable of throwing a javelin up to half a mile (0.8 km) away, and mainly used as a defense weapon on top of fortifications. It consisted of a strong wooden frame and thick ropes that were twisted by a rotating screw and provided the force to throw a javelin with great precision "which no armor could resist." In the Codex Atlanticus, Leonardo drew improved designs of springalds capable of throwing explosive devices instead of javelins.

# FORTRESSES

## the 30-second theory

**3-SECOND SKETCH**
To address the development of powerful new firearms, Leonardo designed new types of fortresses, with polygonal shapes or receding profiles aimed at reducing the impact of incoming missiles.

**3-MINUTE MASTERPIECE**
The type of polygonal fortress studied both by Leonardo and Francesco di Giorgio later became star shape, a form already implicit in some of Leonardo's projects. The fortress surrounding the town of Palmanova in north Italy is one example. It consists of three concentric walls, built in successive historical periods, starting from the end of the 16th century, with the outermost ring built in the Napoleonic era at the beginning of the 19th century.

**Leonardo's plans for fortresses** addressed the major innovation of the day in warfare: firearms. The French army's descent upon Italy, in 1494, equipped with new artillery, had left a powerful impression. In response to this, Leonardo pioneered two new types of fortress: circular fortresses with receding profiles and polygonal, angular fortresses. Fortresses and towers with flattened forms and receding and curved profiles were designed to minimize the area susceptible to attack. The more oblique the surface, the less destructive the impact of incoming projectiles. This innovative concept replaced medieval merlon battlements, which were dominated by straight horizontal or vertical profiles. In Leonardo's most complete project, the general plan of the fortress is a circle with three concentric rings, interspersed with floodable moats. In other projects, he used a polygonal plan, with the walls arranged at sharp angles to allow for a full view of the external field and a more complete firing range. In some cases the two types, curvilinear and polygonal, are combined, as in Leonardo's project for a defensive wall inside a moat.

**RELATED RESEARCH**
See also
THE IDEAL CITY
page 78

BALLISTIC STUDIES
page 106

**3-SECOND BIOGRAPHY**
JACOPO IV APPIANI
1459–1510
Lord of Piombino and other surrounding areas, including the island of Elba, visited by Leonardo, which was one of the many small states that comprised Italy at the time.

**30-SECOND TEXT**
Domenico Laurenza

*Leonardo's plans for a fortress with a low profile and concentric walls look surprisingly modern. Drawing on his understanding of parabolic trajectories, the curved design would have ensured the impact of cannon fire was minimized.*

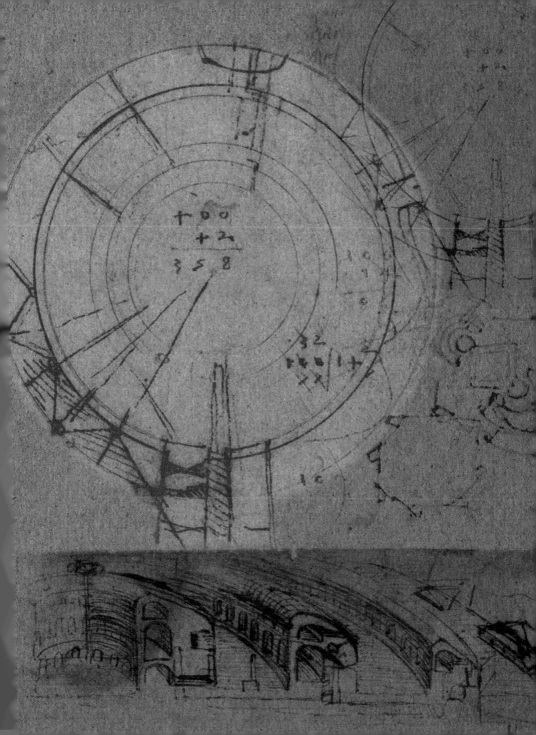

# SPECTACULAR SIEGE MACHINES

## the 30-second theory

### Many of Leonardo's most

famous and spectacular designs for war machines are actually variations upon devices that were already in use from antiquity. In addition to making actual technological improvements to these machines, the originality of his designs consists largely in the unsurpassed quality of the drawings he made, and in the general grandeur of his ideas. These drawings are really the visual counterpart of Leonardo's famous letter to Ludovico Sforza, also known as Ludovico il Moro, Duke of Milan, in which he wrote at length of his abilities as a military engineer. Almost all of these projects date back to Leonardo's first Florentine period or to the early years in Milan after 1482–83. They include his famous drawings of scythed chariots, which, he emphasized, need to be used carefully because they "did not less injury to friends than they did to enemies." Such devices had already been in use since antiquity, as had catapults, although Leonardo developed original loading mechanisms for the latter, incorporating, for example, springs. Similarly, his innovations included, in one case, a design for a gigantic crossbow.

**RELATED RESEARCH**
See also
NAVAL WARFARE
page 108

**3-SECOND SKETCH**
Some of Leonardo's early projects for siege and defense machines, although still bound to traditional warfare, are spectacular in the content and forms of their visual presentation.

**3-MINUTE MASTERPIECE**
As well as spectacular siege engines, some of Leonardo's designs for defense systems are similarly impressive and are also drawn from traditional systems of warfare. These include systems to prevent enemies from scaling the walls, thanks to a long beam pushed against the ladders on the exterior of the walls, or by using a system of rotating blades.

**3-SECOND BIOGRAPHY**
LUDOVICO SFORZA
1452–1508
Fourth son of Francesco Sforza and was Duke of Milan from 1489 until his death. He was patron of Leonardo da Vinci and other artists.

**30-SECOND TEXT**
Domenico Laurenza

*Despite Leonardo's stated aversion to war, his designs for killing machines gave him the patronage of wealthy patrons in rival north Italian cities.*

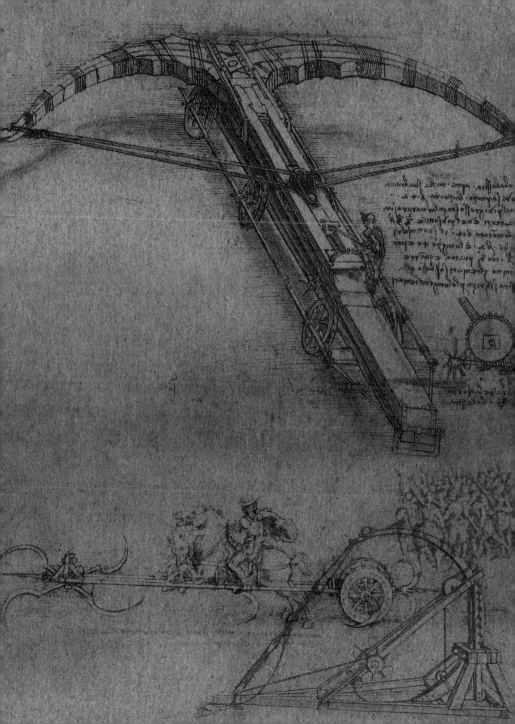

# TANK

## the 30-second theory

### Leonardo applied his dynamic

conception of war to the idea of the tank, an idea which was not entirely new at the time. He mentioned the tank in the famous letter with which he presented himself to Ludovico Sforza, Duke of Milan: "I will make covered vehicles, safe and unassailable, which will penetrate the enemy and their artillery, and there is no host of armed men so great that they would not be broken by them." The most complete example of Leonardo's tank comprised a tortoise-shape chariot that was moved by eight soldiers, who also drove numerous cannons. The chariot was crowned by a conical sighting turret with slits or portholes permitting 360-degree vision of the outside. The tank's main purpose was to directly head and breach the enemy ranks but, as Leonardo notes on the drawing, in order to be effective, it required infantry following its first attack. The large number of light cannons or *scoppietti* arranged around the circumference of the tank recalls another famous Leonardo's project, the *circumfolgore* or multiple-barreled bombard. Although the latter machine was represented as being "open," it was perhaps a covered structure like the tank, and drawings suggest it was probably intended to be mounted on a ship.

**3-SECOND SKETCH**
Leonardo imagined tanks powered by humans or by horses, with numerous cannons attached. They were intended to launch the first attack on the enemy ranks.

**3-MINUTE MASTERPIECE**
In another project, the tank is operated not by men but by horses, placed inside. Leonardo's drawings of the chariot in action depict two successive positions of the chariot launched against the enemy, and include the smoke from the fires and the dust raised by the moving chariot. The use of horses seems unrealistic and has a spectacular dimension shared by many of Leonardo's early projects.

**RELATED RESEARCH**
See also
CANNONS & SPRINGALDS
page 104

NAVAL WARFARE
page 108

**3-SECOND BIOGRAPHY**
LUDOVICO SFORZA
1452–1508
Duke of Milan. He promoted the arts and the culture in his court, but invested much in the art of war. He lost his power in 1499, when the French armies entered in Milan.

**30-SECOND TEXT**
Domenico Laurenza

*Anticipating 20th-century warfare, Leonardo's armored tank may have been his ultimate war machine. Designed to permit simultaneous gunfire from every angle of its conical shell, it would have devastated enemy ranks.*

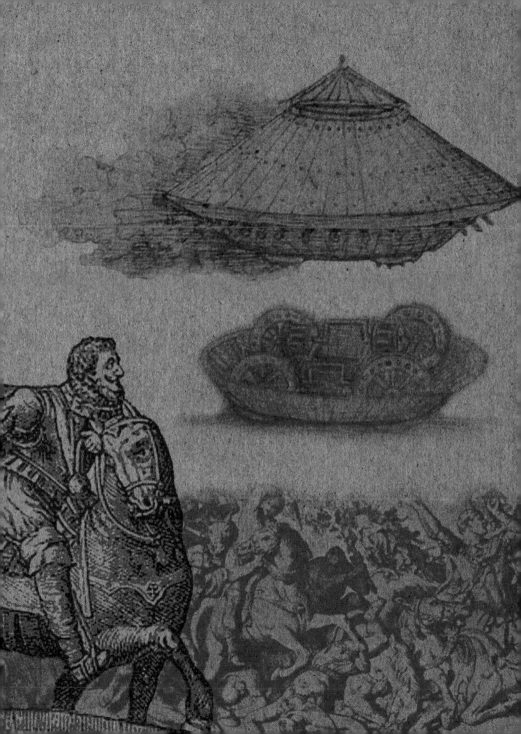

# CANNONS & SPRINGALDS

## the 30-second theory

**3-SECOND SKETCH**
Leonardo's designs for firearms were largely focused on enhancing the rate of fire and creating the possibility of aiming in different directions to cover a wider area.

**3-MINUTE MASTERPIECE**
Leonardo also invented a steam cannon. The barrel of the cannon was preheated to a high temperature, and water poured into it would be converted into steam, the pressure of which would fire a projectile. Leonardo attributed what is almost certainly his original invention to Archimedes, and calls it *architronito*. In so doing, he aimed to present himself as an emulator of the greatest engineer of antiquity, as a "new Archimedes."

### In Leonardo's time, firearms

were transforming the art of war. Leonardo participated in this revolution through his innovative inventions. With his dynamic conception of war, Leonardo was most interested in increasing the mobility and rate of fire of weapons. He therefore focused his efforts mainly on relatively light firearms, such as the "springalds," or medium-caliber artillery pieces smaller than bombards. Thanks to a movable wooden structure, the springald could move both vertically and horizontally. It could also aim in various directions without having to move the entire machine (which was usually held to the ground with ropes due to the strong recoil). His inventions for increasing the rate of fire, for example, his famous designs for multibarrel machine guns, are even more spectacular. One of these was designed to enhance the area covered by the fire. Leonardo proposed a mobile device with a series of barrels or light cannons (*scoppietti*), arranged on a rack with an arched profile, which could be moved both vertically and horizontally. Another example was aimed at increasing the rate of fire, with many multibarrel racks able to rotate and fire rounds in quick succession.

**RELATED RESEARCH**
See also
BALLISTIC STUDIES
page 106

**3-SECOND BIOGRAPHY**
ARCHIMEDES OF SYRACUSE
ca. 287–212 BCE
Greek scientist and engineer and the inventor of famous war machines and hydraulic devices, such as the Archimedes' screw.

**30-SECOND TEXT**
Domenico Laurenza

*Leonardo's sketches for breech-loading artillery, which date from ca. 1480, suggest weaponry with remarkable firepower— the forerunners of the machine gun.*

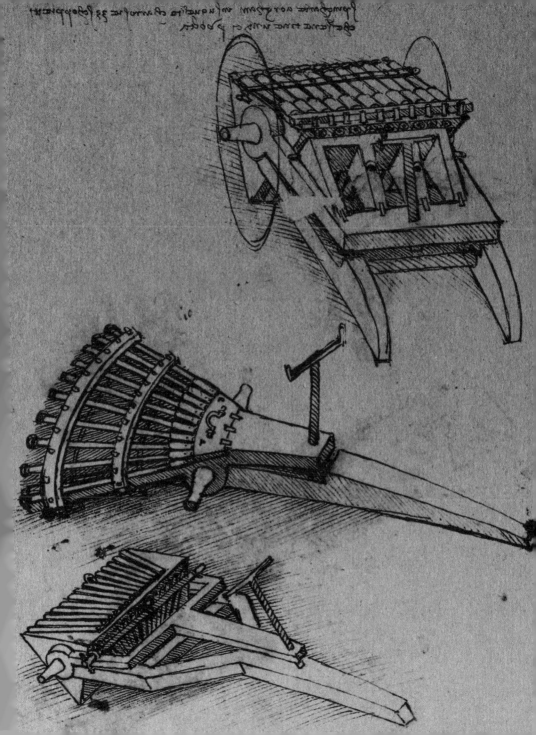

# BALLISTIC STUDIES

## the 30-second theory

### Leonardo's spectacular ballistic

drawings and projects are based on intense scientific studies. In one example, Leonardo demonstrates the relationship between the height of a ball from a bombard and the distance it travels, resulting in magnificent drawings of bombards that could fire projectiles over a wide area. Unlike Leonardo's designs for multibarrel machine guns, the greater rate of fire that these machines achieved resulted not from increasing the number of guns, but by increasing the number of projectiles and their trajectories. Another equally spectacular drawing depicts a large cannonball, whose edges are sewn together, containing several smaller, spherical projectiles. The cannonball explodes shortly after being launched, spreading its contents, which, upon hitting the ground, produce a farther hail of shots. Other designs for aerodynamically shaped ogival projectiles were based on the scientific study of the interference of the air by friction. In some cases, these projectiles had fins, designed to reduce air resistance.

**3-SECOND SKETCH**
Through the scientific study of the dynamics of projectiles, Leonardo sought to increase the area that firearms could cover. He presented these studies in magnificent drawings.

**3-MINUTE MASTERPIECE**
Beyond their technical content, Leonardo's drawings of the trajectories of shots fired from mortars turned ballistic studies into images of great artistic power. Leonardo depicts the trajectories of projectiles as lines of force, creating a series of spectacular arches, like rays emanating from a stellar body. These drawings seem the ballistic counterpart of *the Deluge* drawings created by the artistic transfiguration of his studies on the dynamics of water.

**RELATED RESEARCH**
See also
THE DELUGE
page 32

WATER
page 128

**3-SECOND BIOGRAPHY**
NICCOLÒ FONTANA
TARTAGLIA
1499–1557
Italian mathematician, engineer, and surveyor. First applied mathematics to the investigations of the path of cannonballs.

**30-SECOND TEXT**
Domenico Laurenza

*Leonardo contributed to the study of ballistics in an age when aerodynamic force was not understood and very little was known about gravity. Empirical tests using tiller bows enabled him to calculate the force and distance possible with different setups.*

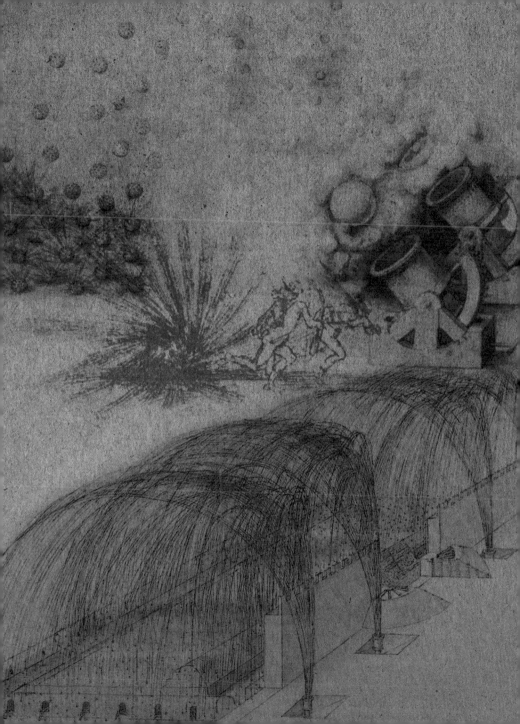

# NAVAL WARFARE

## the 30-second theory

**Leonardo's inventions for naval** war consist of partially encoded texts, fragmentary drawings, and admonitions to himself not to disclose his projects. Because he kept these inventions more secret than others, it is not easy for us to reconstruct them. Leonardo was fully aware of their novelty and even worried about their deadly potential. His projects include a submarine able to ram and damage enemy ships, and ships with a double hull to defend themselves from such attacks. He also devised diving suits made from watertight leather, equipped with ballast for descending and an inflatable bag for ascending. Other ideas included underwater systems to drill and break the bottom of enemy ships operated by divers or submarines. Leonardo also developed methods of staying and moving in water using webbed gloves or air-bags. Just as the flying machine imitates the flight of birds, so these devices imitate the swimming of fishes under and above the water. This strong mimetic and naturalistic conception of the mechanical invention marked Leonardo out from other Renaissance engineers.

**RELATED RESEARCH**
See also
HELICOPTER
page 58

DIVING SUIT
page 70

**3-SECOND SKETCH**
Leonardo invented sophisticated underwater devices, which, while developed largely for use in human warfare, were based on an imitation of the natural world.

**3-MINUTE MASTERPIECE**
In 1500, on behalf of the Republic of Venice, Leonardo projected defensive systems against the Turks along the Isonzo River, in Friuli. A system of mobile dams would be operated using the river's current. Even when intervening in nature, Leonardo's approach is to imitate and work in harmony with nature's own forces, instead of in opposition to them. His approach offers lessons to the engineers of today, who are increasingly concerned with developing eco-friendly technologies.

**3-SECOND BIOGRAPHY**
MARIANO DI JACOPO
(KNOWN AS "TACCOLA")
1382–1458
Along with Renaissance engineers, Taccola projected similar devices, even if much less sophisticated than Leonardo's.

**30-SECOND TEXT**
Domenico Laurenza

*Leonardo made studies in which he compared the hydrodynamic design of fishes to boats, and analyzed the capabilities of the flying fish to move between one element and the other.*

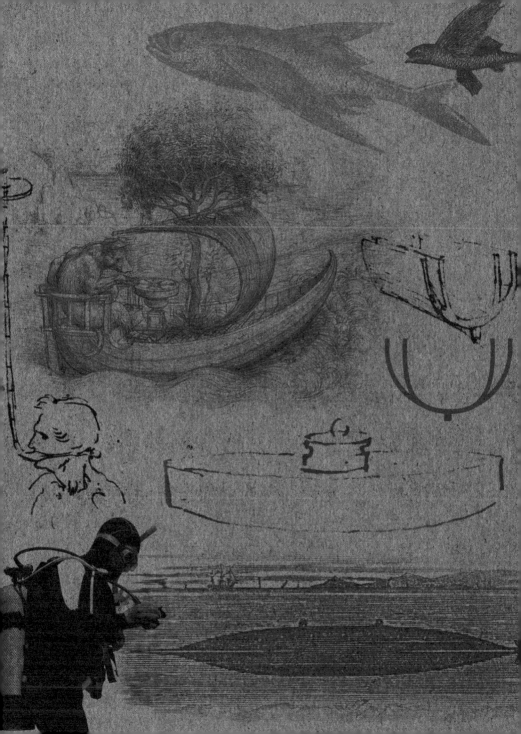

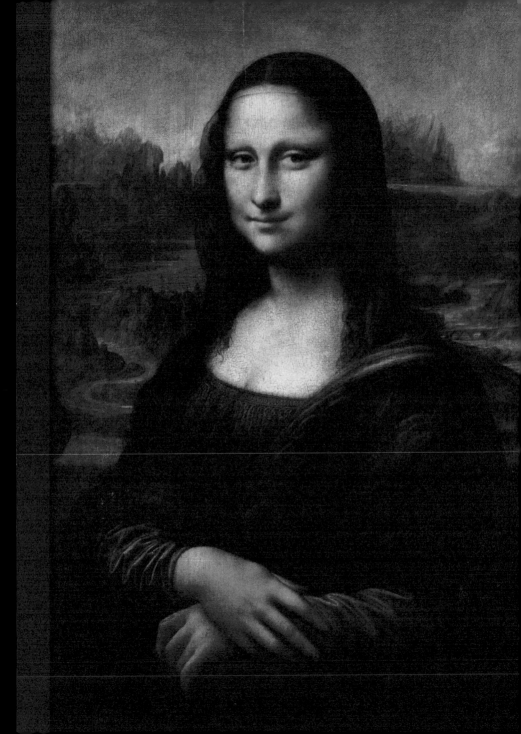

# MONA LISA

**Arguably the most celebrated,** most imitated, and most parodied painting in Western art, the *Mona Lisa* stands as an iconic example of Leonardo's genius, and of Renaissance painting in general. There is no longer much doubt that the sitter was Lisa Gherardini, wife of the Florentine silk merchant Francesco del Giocondo (in a pun on his name she was called "la Gioconda"—"the jovial one"). Leonardo failed to deliver the portrait to his patron. He started the painting in around 1503 (it was known to Raphael when he painted portraits of Agnolo and Maddalena Doni in 1506–7), and it was still in his possession at his death. Leonardo's mature use of translucent glazes suggests that he was still working on the *Mona Lisa* when in Rome, and that possibly he continued to refine it after moving to France in 1516. The panel is unusually large for a female portrait of its date. It includes the hands of the sitter, who is precociously set before a parapet supporting a loggia, below and beyond which stretches an extensive, imaginative landscape backdrop. Here, Leonardo explored the inherent flux of the natural landscape, suggesting its potential for mountainous upheaval, and the seasonal ebbs and flows of its lakes, rivers, and streams.

He sought to show the same inherent life and movement in his sitter, capturing her communicative responses to the observer's presence through direct eye contact. Practicing his own advice to the portrait painter, Leonardo merged his generalizing pictorial light imperceptibly into shadow, hinting at the warm surface vibrancy of Mona Lisa's flesh. Leonardo showed his concern with the "motions of the mind" in his treatment of her facial features, especially her pensive eyes and her mouth that hovers on the brink of a faint, ambiguous smile. He explored these facial movements and expressions by utilizing supremely subtle sfumato shadows around her facial features and along her left temple, cheek, and jaw. These shadows were generated by the repeated application of translucent glazes that also define the gossamer texture of her fine veil. The curls of Mona Lisa's hair and the highlights that stream fluidly from the neckline of her robe offer a parallel sense of incipient movement, envincing Leonardo's analogy between the flow of hair and water. The animation implicit within all aspects of the portrait promotes the sense of mystery that has earned the *Mona Lisa* its unmatched renown.

*Francis Ames-Lewis*

# DISMOUNTABLE CANNONS & DEVICES

## the 30-second theory

### Leonardo's dynamic conception

of war is particularly clear in his designs for dismountable devices, such as bridges or chariots to carry guns. He designed bridges that could be constructed "in short time, in order to flee or follow the enemy." Materials could in some cases be found on the spot. The most refined of these designs includes details of the fixing of the head of the bridge. Leonardo also suggested the use of axes made from strong but thin lumber "like a lance" (he was similarly concerned with the lightness and strength of the materials used in his other projects, such as the flying machine). Alongside systems to speed up the movement of armies, Leonardo devised a way of improving the portability of guns. For example, a sheet of the Codex Atlanticus contains two such designs, in which beams are used both to conceal the gun during transport and as levers to lift it when it was to be fired. Although these projects date back to his early years, they perhaps came in handy later when, in 1502, Leonardo was appointed military engineer to Cesare Borgia, a specialist in surprise attacks.

**RELATED RESEARCH**
See also
THE IDEAL CITY
page 78

BRIDGES
page 84

**3-SECOND SKETCH**
Leonardo designed strong, ingeniously lightweight bridges that were easy to assemble and take apart, and chariots for carrying guns to speed up the movements of armies.

**3-MINUTE MASTERPIECE**
As a military engineer for Cesare Borgia, Leonardo carried out a series of surveys of lands and fortresses in Romagna, the object of Cesare's war of conquest. The most famous and complete of these studies is also one of the masterpieces of Renaissance cartography, one of the first examples of a modern map: a map of the city of Imola, based on accurate measurements of actual distances and their proportionate scale reproduction.

**3-SECOND BIOGRAPHY**
CESARE BORGIA
1475–1507
Also known as "il Valentino," was the son of Borgia Pope Alexander VI. He tried to build a state in central Italy. After his father's death he lost his power and died fighting in Spain.

**30-SECOND TEXT**
Domenico Laurenza

*In an age of military innovation, Leonardo's engineering skills allowed him to design both movable bridges and canons to maintain the element of suprise.*

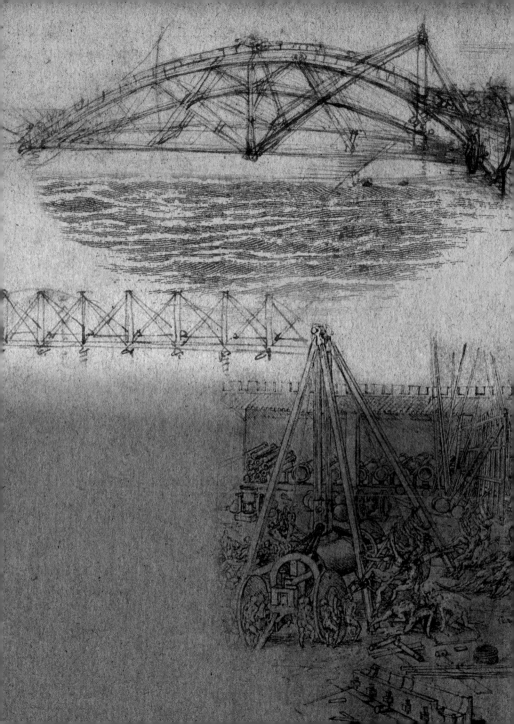

NATURE

**Codex Leicester** The Codex Leicester is an autograph manuscript by Leonardo, composed around 1508–10, and comprising 18 loose double sheets of paper containing his observations on the nature and property of water, as well as other aspects of science and technology. Like most of his manuscripts, it is written in Italian, which is unusual for learned European writings of the time, which were normally in Latin. The Codex Leicester is widely regarded as one of Leonardo's most important scientific notebooks and his only autograph manuscript to be in private hands. Microsoft chairman Bill Gates purchased it from Christie's auction house in New York in 1994 for more than $30 million, making it the most expensive book ever sold. It is put on public display once a year in a different city around the world and all its pages have been scanned and are freely available to view online as digitized files.

**impetus** According to Aristotle's theory of motion, when a mover sets a body in motion, he implants into it a certain impetus, a force enabling a body to move in the direction in which the mover starts it. This explains why a stone moves on after the thrower has ceased moving it. The resistance of the air acts against the impetus and diminishes it until it is not capable of moving the object that then comes to a natural state of rest.

**inertia** Inertia means that if no force acts on an object, then the object at rest will stay at rest, while the already moving object will keep on moving forever along a straight line and at a constant speed.

**macrocosm and microcosm** An ancient Greek theory of seeing the same patterns at all levels of the cosmos, from the largest, the arrangement of the planets (macrocosm), to the smallest, that of the most minute components of the human body (microcosm).

**percussion** In his Manuscript A, kept at the library of the French Institute in Paris, Leonardo defines percussion as "motion interrupted by a resisting object" or "the terminus of incident motion and the beginning of reflected motion achieved in an indivisible speed, time, and position."

**prime mover (*primum mobile*)** Aristotle believed that all movement depends on there being a mover, and that behind every movement there must be a chain of events that leads back to something that moves but is itself unmoved and is the purpose of the movement of all things. To Aristotle, movement meant more than just travel from one place to another but included every change in a universe in a state of flux. This he referred to as the prime mover, a being with everlasting life, continually applying the force, and in his *Metaphysics*, Aristotle also calls this being "God."

# THE PRIME MOVER

## the 30-second theory

### 3-SECOND SKETCH
Leonardo strove to understand the laws of nature, in particular how motion can be explained. The concept of the prime mover was considered to be behind all movement in the universe.

### 3-MINUTE MASTERPIECE
In the Codex Leicester, Leonardo drew small figures of men performing various actions corresponding to the discussion he develops on the pages. Folio 8A is devoted to a complex discussion of the principles of percussion and impetus in moving bodies and fluids. Leonardo draws movement by demonstrating both impetus and percussion. There is a sense that his figures move with a perceptible intent.

### Leonardo wrote extensively

about movement, trying to understand what lies behind such a fundamental law of nature, observable all around us. His investigations led him to believe in the Aristotelian concept of the prime mover (*primum mobile*) as the fundamental force that underlies all of nature. This belief was popular at the time, and had also infused late-Medieval thought. The movement of the planets was attributed to divine agency, that is, the divine act of setting all things in motion. The prime mover was the original unmoved force beyond the limits of changeable and unpredictable events that were linked to nature, to its course, its rules, and its innumerable expressions. Furthermore, "impetus," according to Medieval philosophy, was behind the forces that moved and kept all things in the world in motion. The concepts of impetus and of the prime mover were linked to the correspondence between the macrocosm and the microcosm. For instance he saw, within this framework, that the water that emerges exuberantly from mountain springs, in an act of relentless motion, was comparable to the sap that oozes from the top of a severed vine, or the blood that pours out of a cut vessel in the heads of humans and animals.

### RELATED RESEARCH
See also
MACROCOSM & MICROCOSM
page 124

WATER
page 128

### 3-SECOND BIOGRAPHIES
JEAN BURIDAN
(JOHANNES BURIDANUS)
ca. 1300–after 1358
A French priest who developed the concept of impetus, the first step toward the modern concept of inertia, and an important development in the history of medieval science.

BLASIUS PELACANI DA PARMA
ca. 1365–1416
Italian philosopher, astrologer, and mathematician who developed theories on the laws of motion and also wrote on the natural philosophy of Aristotle.

### 30-SECOND TEXT
Marina Wallace

*Using Aristotle's ideas as a backdrop for his own, Leonardo looked for connections between the nature in front of him and the universe as a whole.*

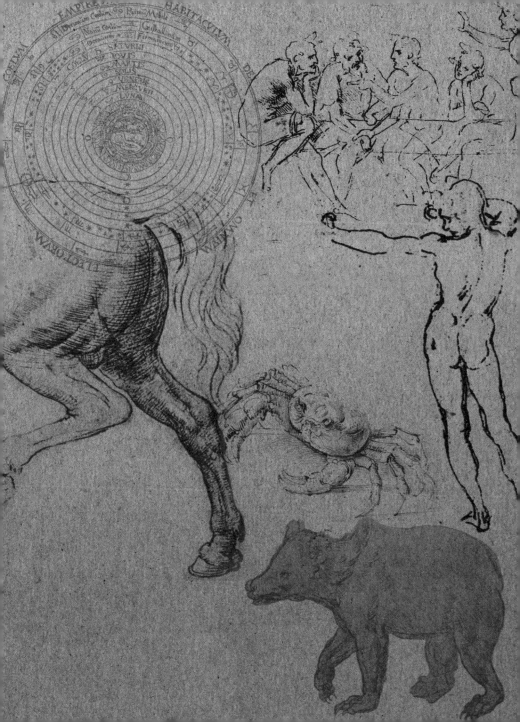

# THE BODY OF THE EARTH

## the 30-second theory

**Leonardo thought that the body** of the earth was comparable to that of man, not in the sense that they looked similar, but in that the principles of organization in each worked in parallel ways at a very profound level. The depth of his argument and of his investigations cannot be sufficiently stressed. In his vision of the world, Leonardo saw the earth as a living planet, with all of its elements in a constant state of flux. His mind worked best when applied in a way that developed analogies to explain the mysteries of the universe. For example, he compared the trachea in the human body with the branching of trees. After he dissected a one hundred-year-old man in 1508, Leonardo wrote: "The earth has a vegetative spirit in that its flesh is the soil, its bones are the configurations of the interlinked rocks of which the mountains are composed ... and its blood is the water in the veins; the lake of blood that lies within the heart is the oceanic sea, and its breathing is the increase and decrease of the blood during its pulsing, just as in the sea is the flux and reflux of water."

**3-SECOND SKETCH**
Leonardo made numerous analogies between the body of the earth and that of man.

**3-MINUTE MASTERPIECE**
Leonardo mapped the body of the earth as he mapped the human body, drawing actual landscapes with such accuracy that the topography can be recognized today. A number of Tuscan scenes can be identified in his drawings, like one particular site in Lombardy on the Adda River near Villa Melzi, the family home of his pupil Francesco. We can recognize a little arched bridge over a stream. The site of the background of the *Mona Lisa* has also been identified by some as the Burano bridge in Tuscany.

**RELATED RESEARCH**
See also
MACROCOSM & MICROCOSM
page 124

WATER
page 128

**3-SECOND BIOGRAPHIES**
PLINY
23–79 CE
Wrote in his *Natural History* that it is well known that heavy rains follow transitions of Saturn.

PROCLUS LYCAEUS
412–485 CE
Greek Neoplatonist philosopher who recorded the beliefs of the Pythagoreans: "Again, in the heavens, Ares is fire, Jupiter air, Kronos water

**30-SECOND TEXT**
Marina Wallace

*Leonardo asserted that "The idea appeals to me that the Earth is governed by nature and is much like the system of our own bodies..."*

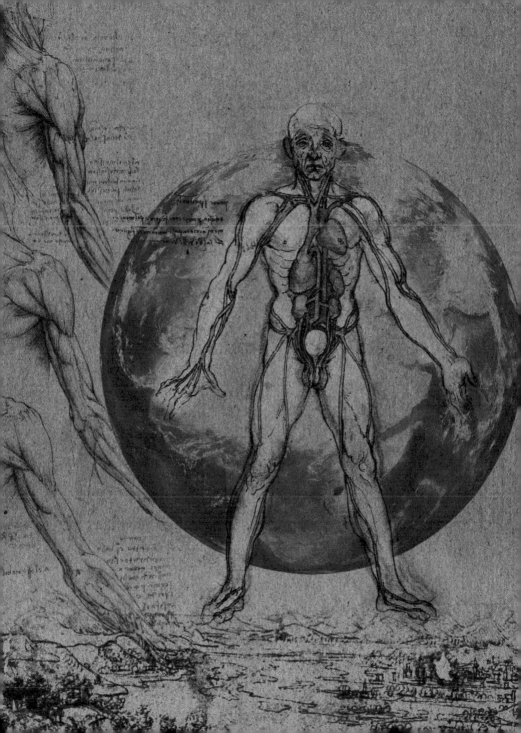

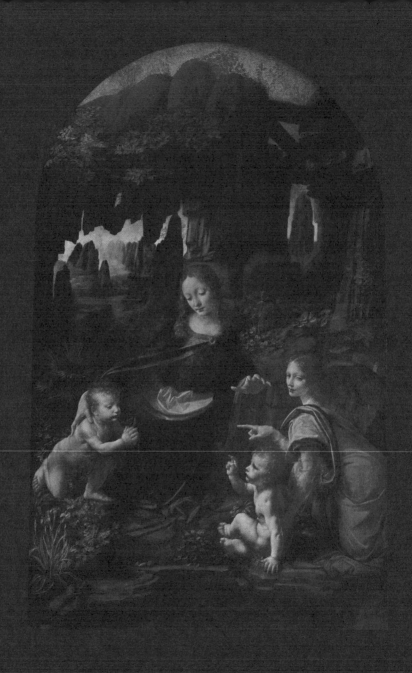

# VIRGIN OF THE ROCKS

## Two versions survive of the

*Virgin of the Rocks*, the early masterpiece of Leonardo's first Milanese period. Although we still have many documents relating to it, the project's history cannot be definitively reconstructed. The first version (now in the Louvre, in Paris, illustrated here) was commissioned on April 25, 1483, by the Milanese Confraternity of the Immaculate Conception as the main panel for the altarpiece in their new chapel in San Francesco Grande. Leonardo and his assistants, Ambrogio and Evangelista da Predis, regarded the Confraternity's 1489 valuation of the panel as too low, however, and they probably sold it, perhaps to Ludovico Sforza, Duke of Milan. Infrared reflectograms show that the second version (in the National Gallery, in London) originally employed a different composition, but probably by the time he left Milan late in 1499, Leonardo had started to paint a close copy of the first version. Although not completely finished, this was installed in the altarpiece frame by October 23, 1508, the date of the final payment, over 25 years after the work was originally commissioned. In the Paris panel, the figures' gestures and facial expressions exemplify Leonardo's concern to develop a narrative exchange among them. This is choreographed by the Virgin Mary's central, dominating presence, and is aided by St. John the Baptist, whose inclusion was not required by the original contract. Leonardo's subtle handling of light reinforces this by spotlighting hands and expressive faces, which emerge brightly from the surrounding shadows. Light is controlled by the group's enclosure within a rocky grotto; the figures are softly, selectively lit from the left. This carefully directed lighting generates imperceptible changes of tonality to develop the delicate sfumato modeling of form and relief. The plants in the foreground are also strongly lit to emphasize their symbolic values as attributes of the Virgin. The saturated color of the fully lit foreground forms gives way to a progressively more muted tonality as the scene recedes toward the jagged rocks of the distant landscape. Leonardo's revolutionary handling of light and color is somewhat masked by the distorting effects of layers of age-old, yellowed varnish. Cleaning and conservation might produce as much of a revelation as has the recent restoration of the *Last Supper*. The second version of the *Virgin of the Rocks*, probably started in the mid-1490s, was painted principally by Leonardo, with the assistance perhaps of Ambrogio da Predis, notably in the foreground and in the rocky backdrop.

*Francis Ames-Lewis*

# MACROCOSM & MICROCOSM

## the 30-second theory

The microcosm–macrocosm analogy was important in Renaissance thought, given that man was considered to be at the center of the universe. Macrocosm and microcosm were already part of ancient Greek thought. Leonardo himself referred in his manuscripts to this ancient analogy, in particular when describing the human body, and extended this theory to the whole of nature. He relentlessly observed details in nature, and drew parallels with similar details in the body of animals and man. He stated in his notes that the design of nature must be respected in all its marvelous forms. All designs have a particular function, from the majestic force of the deluge to the minute valve in the heart. Among the many comparisons he made between the body of man and that of the earth, he envisages the trachea in relation to the branches of trees. There is a good deal of crossover between Leonardo's concept of the body of the earth, and that of microcosm and macrocosm. He writes: "the total amount of air that enters the trachea is equal to that in the number of stages generated from its branching like ... a tree in which each year the total estimated size of branches when added together equals the size of the trunk of the tree."

**RELATED RESEARCH**
See also
THE BODY OF THE EARTH
page 120

DISSECTION & VISUALIZATION
page 138

**3-SECOND SKETCH**
For Leonardo, the macrocosm—that is the universe as a whole—was mirrored in the microcosm, or "lesser world" of the human body.

**3-MINUTE MASTERPIECE**
Leonardo strove to represent a synthesis of many systems in one image. He came close to doing this with the extraordinary drawing of the respiratory, vascular, and urinogenital systems of a woman. This "portrait" of the inside organs of a woman must be seen as an attempt to connect with the concept of the microcosm, the human body being a representation of what Leonardo called a "lesser world," in the sense of a smaller world in comparison with the magnitude of the universe.

**3-SECOND BIOGRAPHY**
PICO DELLA MIRANDOLA
1463–1494
Italian Renaissance philosophe and author of the *Oration on the Dignity of Man*, considere to be the "Manifesto of the Renaissance."

**30-SECOND TEXT**
Marina Wallace

*Interconnection was key to Leonardo; he identified a common mechanism in the myriad vessels of the human respiratory system, the branching nature of a tree, and the network of rivers and their tributaries draining the earth.*

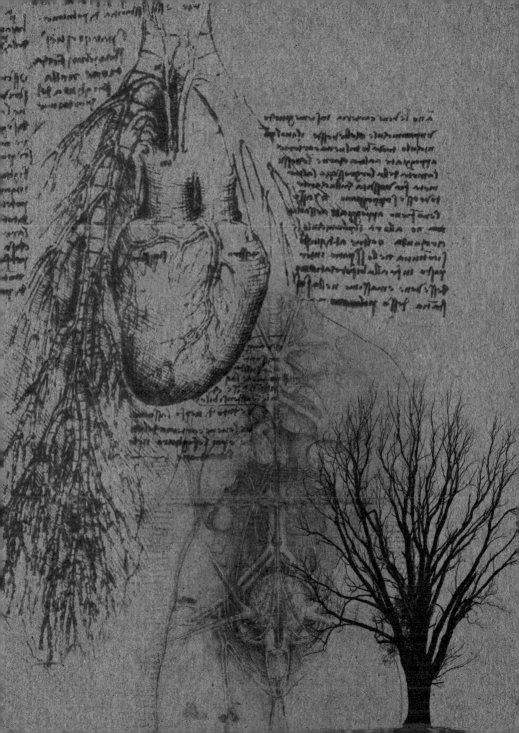

# NECESSITY & EXPERIENCE

## the 30-second theory

**3-SECOND SKETCH**
Form follows function in nature. Necessity and geometry work hand in hand to produce all natural forms where nothing is in excess or in defect.

**3-SECOND SKETCH**
Form follows function in nature. Necessity and geometry work hand in hand to produce all natural forms where nothing is in excess or in defect.

**3-MINUTE MASTERPIECE**
There are 600 of Leonardo's drawings kept in the Royal Library, in Windsor, UK. Many of them are beautiful studies of plants and animals that he used for his paintings. They reveal the extreme precision with which Leonardo studied plants, leaves, and flowers in the belief that nature's design manifested perfection and economy at the same time. Everything has its place in nature and Leonardo's drawings are an eloquent demonstration of this.

According to Leonardo's view of the world, nature is "mistress of the masters." In addition to this, and in order to completely comprehend cause and effect in the superb design of nature, he identified one supreme principle that overrides all others. This is the principle of necessity (*necessitá*) that incorporates other fundamental concepts, such as that of the direct correspondence between form and function, with no insufficiency or redundancy. According to the principle of necessity, every force expends itself in the most direct way available to it. The idea of necessity prescribes that the simplest design will be the most efficient to achieve a given end. Leonardo felt that such stringent principles should be respected by man and should prevail over any contrivance or artificial endeavor. The whole of nature itself, according to Leonardo, is governed by its fundamental principle of necessity. Leonardo concluded that a detailed geometrical analysis was required in order to understand the way that shells assume a heliacal shape, how leaves and petals originate from stems, and the reason why a heart valve works with perfect economy.

**RELATED RESEARCH**
See also
PLANE GEOMETRY
page 46

**3-SECOND BIOGRAPHY**
GALILEO GALILEI
1564–1642
Italian physicist, mathemat and astronomer whose observations and investig were also based on conce similar to Leonardo's in re to the field of necessity a experience.

**30-SECOND TEXT**
Marina Wallace

*Leonardo believed that the universal architecture of necessity was geometry, stating "necessity is the mistress and teache of nature, the curb, rule, and the theme*

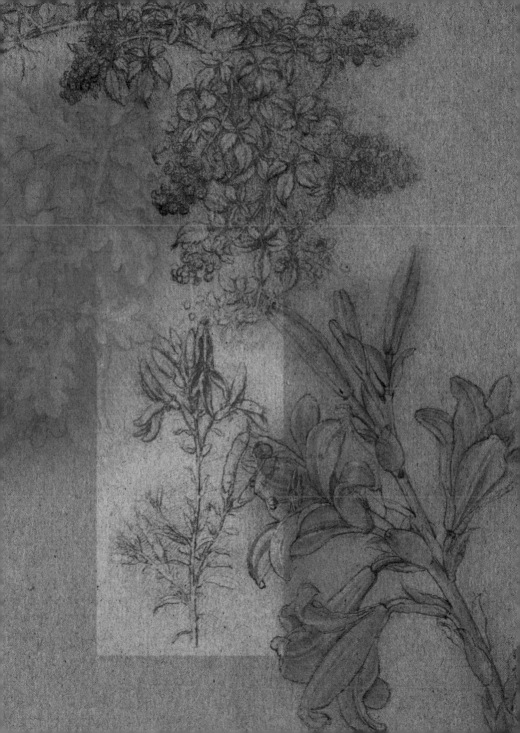

# WATER

## the 30-second theory

The most visible and dynamic of the elements—water—provided a means of understanding the forces that drove all the elements into action. Leonardo studied water relentlessly. He compared the flow of water to the flow of air, and water waves to sound and light. He proposed that the earth was contained within a "watery sphere," which he illustrated as a triangle in a circle with its angles protruding from the circumference. He described the principle as a pyramid embedded within a sphere of water. The projecting elements represented landmasses and mountains. In order to illustrate why the earth remained spherical, he provided instructions to submerge a cube of lead "the size of a grain of millet attached to a string in a drop of water." The drop would not lose any of its original roundness, although it would be "increased by an equal amount to the size of the cube shut within it." However, one problem bothered Leonardo all his life, as it did his predecessors and contemporaries. Water emerged with great force from the body of the earth. But how could this occur if the upper level of the water was far below, in the surface of the seas?

**3-SECOND SKETCH**
Leonardo visualized the world as criss-crossed by a circulatory system of channels comparable to the body of man, through parallel laws of nature that he saw governing the body of earth.

**3-MINUTE MASTERPIECE**
Ludovico Sforza set up a model farm near Vigevano, not far from Milan, with the plan of creating a state-of-the-art irrigation system. Leonardo visited the farm and made observations of the mechanisms of gates and sluices, working out ways in which he would have improved the plan. He made a small but telling sketch of a series of water steps. Another wonderful drawing by Leonardo relating to controlling waterways is of a schematic plan of Florence and the Arno River.

**RELATED RESEARCH**
See also
CARTOGRAPHY & CANALS
page 82

THE BODY OF THE EARTH
page 120

**3-SECOND BIOGRAPHY**
LUDOVICO SFORZA
1452–1508
Fourth son of Francesco Sforza. He was Duke of Milan from 1489 until his death. He was patron of Leonardo da Vinci and other artists.

**30-SECOND TEXT**
Marina Wallace

*Water was the single theme that preoccupied Leonardo throughout his life and was the subject of some of his most extraordinary drawings and ingenious practical schemes, including one for diverting the River Arno, drawn ca. 1504.*

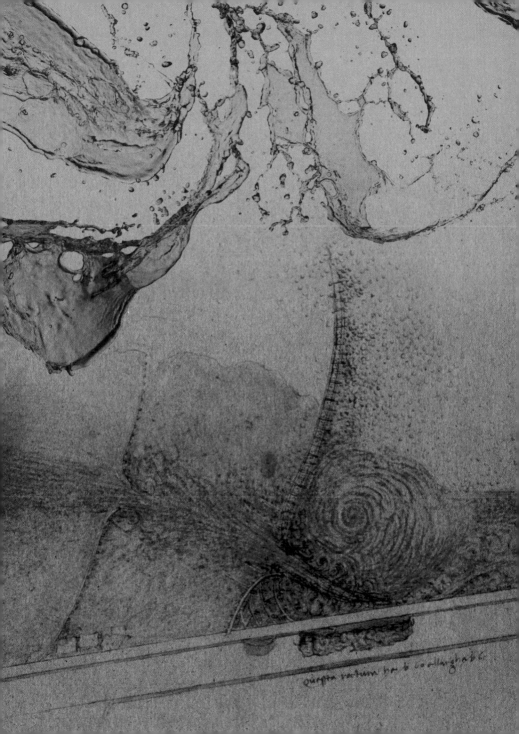

# THE VORTEX

## the 30-second theory

### 3-SECOND SKETCH
Leonardo was fascinated by turbulent motion. Vortex motion was especially powerful, since a vortex speeds up toward its center. He created compelling images of water in spiral motion.

### 3-MINUTE MASTERPIECE
Leonardo's extraordinary drawings of the heart valves, to study the mechanism of the turbulence of the blood, were made around 1513 from direct observation, dissecting the heart of an ox. He envisaged the three cusps in a drawing that is exemplary in its clarity. He also devised graphic models for the blood flow, drawing comparisons to spirals in Ionic capitals.

**The phenomenon of the vortex** was, for Leonardo, a great source of inspiration. He observed vortices above all in water, and drew them in ways that resembled the spiraling curls of hair or the leaves of plants that often appear in his art. His belief in the interconnectedness of all natural forms and the all-governing creative power of nature is strongly manifest in his observation of vortices. The spiral form present in vortices must have been an added attraction for Leonardo, who was intrigued by the shape of spiral staircases and spirals on shells. In analyzing the form of the vortex, Leonardo identified two parts: the primary direction of motion in a straight line, and the revolving motion resulting from the element encountering its own mass. He wrote: "*Note the motion of the surface of the water, which conforms to that of the hair, which has two motions, one of which responds to the weight of the strands of the hair and the other to the direction of the curls.*" Leonardo held the title "Master of Water," a position that allowed him to bring theory and practice together. When consulted by the authorities in Florence and Venice he aimed to demonstrate why vortices were consuming the banks of the rivers and were a threat to the safety of the cities.

### RELATED RESEARCH
See also
THE DELUGE
page 32

WATER
page 128

### 3-SECOND BIOGRAPH
FABRICIUS OF PADUA
1537–1619
An anatomist and surgeon investigated, among other things, the flow of blood in the veins.

WILLIAM HARVEY
1578–1657
English physician, was the to describe how the heart pumps blood into the body (blood circulation).

### 30-SECOND TEXT
Marina Wallace

*The vortices Leonard observed were omni-present: in turbulent water, in gentle rive ripples, in the majest waves of the sea, in curling leaves, tendr human hair, and ma other natural elemen*

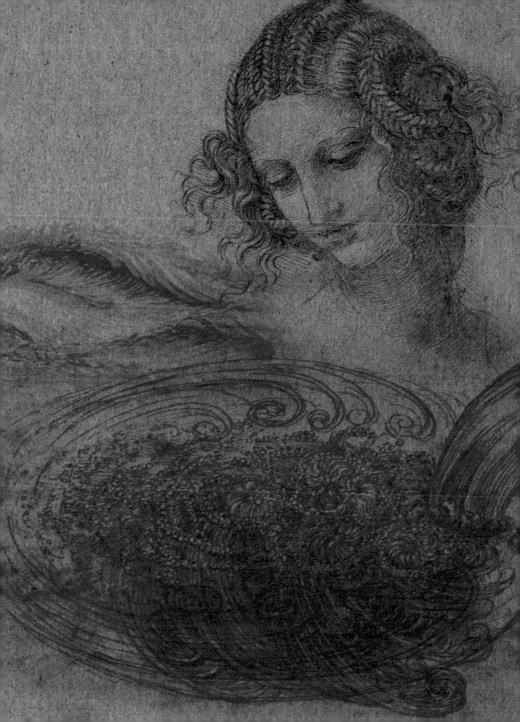

# FORCES OF NATURE

## the 30-second theory

**The most dramatic example of** Leonardo's study of the body of the earth involves his extensive search for the origin of fossils and the biblical Flood. The interpretation of Genesis was that when God divided the earth and the waters, he set them up for all time as they are arranged now. However, ancient philosophers recorded the presence of shells from marine creatures high on mountainsides and other geological evidence of huge transformations in the topographies of land and sea, and for many classical authors, such as Seneca, it was apparent that the positions of the earth and waters could not always have been as they were. One Medieval explanation was that "fossils" were "sports" of nature, arising as the result of some kind of magic deposit or astrological activity. Needless to say, Leonardo dismissed this theory. He also had great problems with the idea that the strata of shells testified to the reality of the biblical account of the total inundation of the earth and the preservation of God's creatures aboard Noah's ark. He records tales of earthquakes swallowing seas and forests sinking down into the earth accompanied by torrential floods. Leonardo's representation of the invisible forces of nature is strikingly modern, both in terms of its graphic ingenuity and expressive power.

### 3-SECOND SKETCH
Leonardo was fascinated by the forces of nature. He observed and described the destructive power of earthquakes, torrential floods, and strong winds. He questioned Medieval theories of the origin of fossils and challenged the conventional explanations of the biblical Deluge.

### 3-MINUTE MASTERPIECE
Leonardo would have been a moviemaker, had the medium been discovered in his time. His drawings have a cinematic effect. As part of his interest in the forces of nature, he drew an exploding mountain, drafting a series of echoing forms, detailing stratified rocks that fan out in the explosion. A potent suggestion of motion is often visible in many other drawings.

**RELATED RESEARCH**
See also
THE BODY OF THE EARTH
page 120

WATER
page 128

### 3-SECOND BIOGRAPHY
PLINY
23–79 CE
Roman natural philosopher and author of *Naturalis Historia*, who died while observing the eruption of Vesuvius over Pompeii in 79 CE.

### 30-SECOND TEXT
Marina Wallace

*"The Deluge" drawing may be Leonardo's apocalyptic vision of a landscape beset by catastrophe of biblical magnitude, in which uncontrolled currents of air, cloud, rain, and water are visible yet merge into all-consuming universal turbulence*

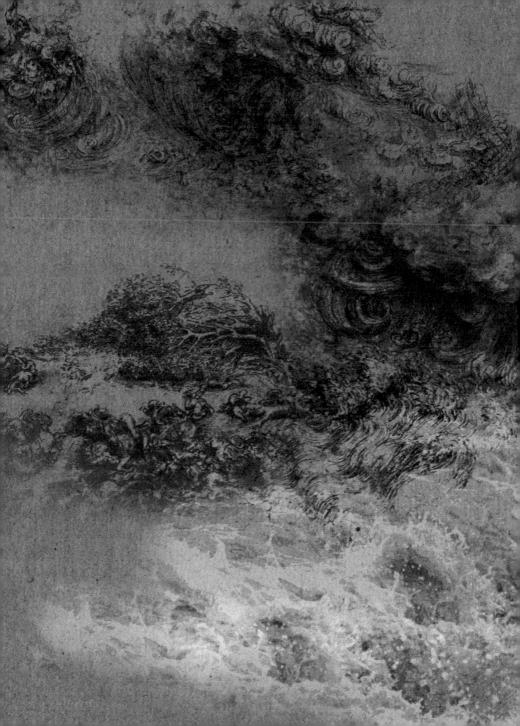

# ANATOMY & ANATOMICAL STUDIES

## ANATOMY & ANATOMICAL STUDIES
GLOSSARY

**Achilles tendon** In his studies of the knee and ankle joints, Leonardo consistently resorts to the analogy with the lever. He describes the Achilles tendon as a lever that raises the heel, using the fore part of the foot as a fulcrum, and measures the tendon's exact strength by applying the laws of levers.

**anthropometry** The study of the measurements and proportions of the human body. In the text around his drawing of the *Vitruvian Man*, Leonardo lists a number of his observations regarding the relationship of various body measurements, such as "the length of the outspread arms is equal to the height of a man."

**calcaneus** The heel bone, the largest bone in the human foot, to which the Achilles tendon is attached.

**the common sense** According to Aristotle, the five senses (sight, hearing, smell, taste, and touch) meet in a "higher unit" that analyzes and processes the perception of the senses into a whole. In his *De Anima* he speaks of this midpoint in our mind as "*koine aisthesis*," or common sense.

**comparative anatomy** Leonardo's interest in anatomy wasn't limited to studying the human body; he also dissected horses, cows, and dogs so he could compare their structure to that of the human body. He produced many drawings of the same limb in the same position, one human, the other belonging to an animal, in order to make a side-by-side comparison.

**dissection** The careful and methodical deconstruction of a human body performed by cutting it up to study its internal components. During the winter of 1510–11, Leonardo performed no fewer than 20 dissections of human bodies with Marcantonio della Torre, the Professor of Anatomy at the University of Pavia. In his Anatomical Manuscript A, Leonardo described in more than 13,000 words and 240 drawings the human body with startling accuracy. Comparison with modern medical imaging shows how truly ground-breaking Leonardo's investigations were.

**Hippocratic medicine** Hippocrates (460–370 BCE) was a Greek physician now acknowledged as "the father of modern medicine." He was the first to show that disease was a natural process and that the symptoms of a disease were due to the natural defense reactions of the body. He realized that the human body functioned as a whole (*"physis"*) and should be treated as such through a holistic approach emphasizing the importance of diet and exercise.

**humors** The four humors of Greek medicine are the four vital fluids that are essential to the proper functioning of the body and influence the mind, thoughts, and emotions. When in balance (*eucresia*), there is health, when unbalanced (*discresia*), there is disease. They are created by the liver and consist of blood, phlegm, yellow bile, and black bile.

**physiognomy** The relationship between a person's physical appearance, especially their face, and their character. An ancient Greek text called *Physiognomika* ("the art of judging a person's character by his or her face"), attributed to philosopher Aristotle, proposed a system for interpreting facial features. Leonardo didn't care for this theory without scientific foundation, but still considered lines caused by facial expressions to be indicative of a person's character.

**vascular system** (circulatory system) The vascular system consists of the heart, arteries, veins, and capillaries. Blood is pumped by the heart and brings oxygen and nutrients to the cells of the body's tissues and organs. It then returns to the heart after being enriched with oxygen by the lungs.

**the Windsor Collection** The Royal Library at Windsor Castle, UK, houses more than 600 drawings by Leonardo, making it one of the world's largest collections. They span the major periods in Leonardo's life and include portraits and anatomical studies as well as botanical sketches.

# DISSECTION & VISUALIZATION

## the 30-second theory

Taking forward the tradition that had endured since the 14th century, Leonardo stressed the importance of dissection, and performed dozens of dissections of human bodies. These were not carried out in secret places, as the popular myth would suggest, but in hospitals, such as those of Santa Maria Nuova in Florence or Santa Maria della Consolazione in Rome. The absence of modern preservatives meant that dissections took place in winter, when corpses stayed in good condition for longer. In Florence, he dissected the body of an old man and, simultaneously, a child of two years to study how bodies change over time. In Rome he performed dissections to understand the relationship between mother and fetus. Between Milan and Pavia he conducted more systematic dissections, in collaboration with the young anatomist Marcantonio della Torre. Dissection was an analytical process. To understand the form of a vein required the destruction of the surrounding organs and tissues. Hence, there was a need for images that could depict not only the individual parts of the body in isolation, but also in their natural positions relative to other body parts. Leonardo was able to capture this, and the visual complexity of his anatomical images remained unsurpassed for centuries.

**3-SECOND SKETCH**
Through his extremely detailed visual representations of anatomy, Leonardo could have helped to transform the study of medicine, which had relied largely on written descriptions.

**3-MINUTE MASTERPIECE**
In 1543, Andreas Vesalius published the *De Humani Corporis Fabrica*, which was considered the editorial masterpiece of Renaissance anatomy, and was particularly celebrated for its beautiful plates. These anatomical images were less complex than those of Leonardo but, being published, were widely seen by other scientists, whereas those of Leonardo were not. Vesalius, besides being a great anatomist, was also a skilled manager of his work, unlike Leonardo.

**RELATED RESEARCH**
See also
LINEAR PERSPECTIVE
page 38

**3-SECOND BIOGRAPHI**
MONDINO DE' LIUZZI
ca. 1270–1326
Italian anatomist and professor of surgery.

MARCANTONIO DELLA TOR
1481–1511
Italian anatomist.

ANDREA VESALIUS
1514–1564
Flemish anatomist, physicia and surgeon.

**30-SECOND TEXT**
Domenico Laurenza

*Leonardo's myriad contributions to the study of anatomy extended to method, language, and conte and represent some of the most accurate studies ever made.*

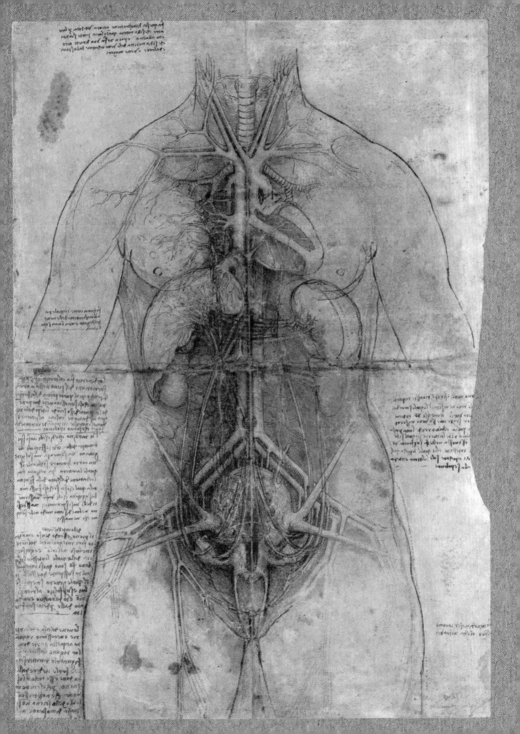

# THE BODY MACHINE

## the 30-second theory

### Leonardo often applied concepts

to the human and animal body that he had developed in his studies on machines and their component parts. He subjected the human body to an intense static and dynamic analysis, studying the body as a whole and the limbs as systems of levers rotating around their fulcrums. For example, in the lower leg, he analyzed the lifting motion of the calcaneus close to the Achilles tendon, a system that works like a lever with its fulcrum in the front part of the foot. He compared the spine to the mast of a ship, supported by muscles and ribs, as well as the rigging securing the mast. The production of heat by the heart led him to compare it to a furnace. He also compared the way the chest lifts during breathing to a weight that raises a rod using a pulley. This static and dynamic analysis was also applied to the movements of head, limbs, and torso in the body as a whole, in studies that have been almost completely lost but of which we have an indirect testimony in the drawings contained in the Code Huygens, now in the Morgan Library in New York.

**3-SECOND SKETCH**
Leonardo studied the animal body using principles of mechanics, analyzing its movement in terms of levers, fulcrums, and other similar engineering concepts.

**3-MINUTE MASTERPIECE**
Leonardo designed a humanoid robot and a mechanical lion. The robot was able to walk, move its arms, and open its mouth. Even the mechanical lion, designed for a feast held in France in 1515 in honor of Francis I, was able to walk a few steps, to sit up on its hind legs, moving its tail, and lastly open its chest to show a bouquet of lilies.

**RELATED RESEARCH**
See also
SELF-PROPELLED CART
page 72

**3-SECOND BIOGRAPHY**
FRANCIS I
1494–1547
French monarch and Leonardo's final patriarch.

**30-SECOND TEXT**
Domenico Laurenza

*Leonardo's efforts to depict in two dimensions the complex range of motion of limbs and joints, are known from present-day computer animated drawings to be representations of astonishing accuracy.*

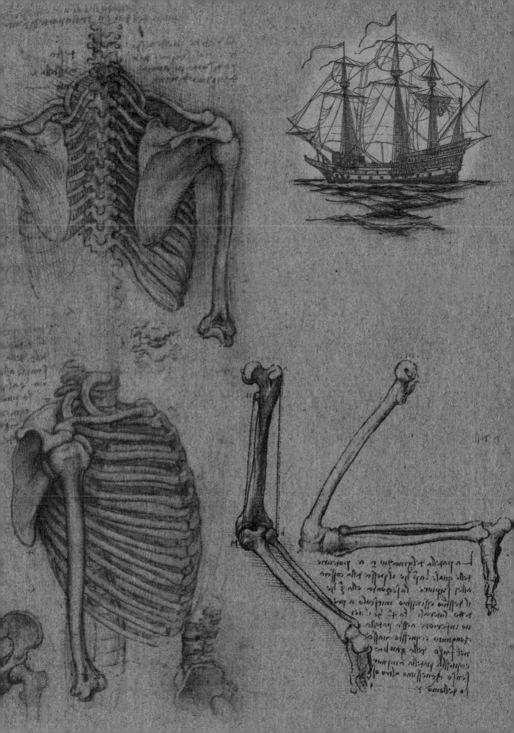

# THE FLUIDS OF THE BODY

## the 30-second theory

### 3-SECOND SKETCH
Leonardo studied all body fluids, especially blood and its movements in the cavities of the heart, which he discovered to be a muscle.

### 3-MINUTE MASTERPIECE
On the basis of the ancient analogy between macrocosm and microcosm, or body of the earth and body of man, Leonardo compared the heart to the ocean and the vessels to the rivers. This analogy, which Leonard himself revisited critically in later life, extended to architecture and urban planning, as in the case of the project for a city with input channels for goods and output channels for the expulsion of waste.

In Leonardo's day, fluids were of paramount importance within medicine. At the time, it was believed not only that the urinary, respiratory, and intestinal systems were fluid-based, but even nerve impulses were conceived of as an aerial fluid that "wafted" in the nerves from the brain. Moreover, Hippocratic medicine had been based on the notion of four fluid "humors"—four temperaments. Leonardo's most significant contribution was on the blood and vascular system. He discovered that the heart is a muscle and that, therefore, life is a matter of strength. He also studied how the blood is subjected to intense back and forth movements in the chambers of the heart, which, by friction, make it increasingly finer, transforming it from venous into arterial blood, which is less dense, full of warmth, and contains air. Circulated through the arteries, this blood enlivened the whole body and its organs. Leonardo did not discover the circulation of blood, but his studies of the heart are, from an anatomical point of view, more detailed than those of William Harvey, recognized as the discoverer of blood circulation. Some of Leonardo's studies are based on human dissections. Others, made while he was a guest of his pupil Francesco Melzi in Vaprio d'Adda (ca. 1513), are based on animal dissections.

### RELATED RESEARCH
See also
THE IDEAL CITY
page 78

THE BODY OF THE EARTH
page 120

### 3-SECOND BIOGRAPHIES
FRANCESCO MELZI
ca. 1492–1570
Painter, nobleman, and a pupil of Leonardo.

WILLIAM HARVEY
1578–1657
Physician and anatomist and the first to describe in detail the circulation of blood.

### 30-SECOND TEXT
Domenico Laurenza

*Deducing the functional anatomy of the heart led Leonardo to revisit the Renaissance view of fluids of the body operating as part of the macrocosm/microcosm continuum in which human biology was determined by the same principles observed in the natural world.*

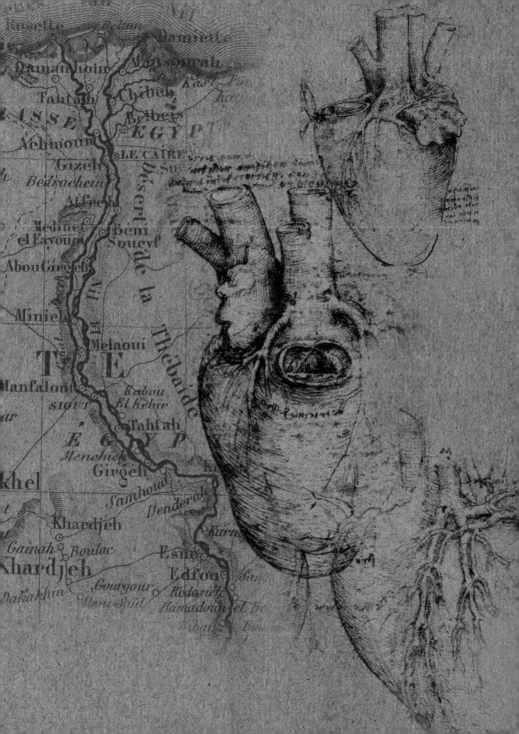

# SEARCHING FOR THE SOUL

## the 30-second theory

**3-SECOND SKETCH**
Leonardo attempted
to locate the anatomical
seat of the intellective soul
at the center of the skull.

**3-MINUTE MASTERPIECE**
Leonardo also studied
the seat of the common
sense in the cavities or
"ventricles" of the brain.
In one of his experiments
he injected the brain of
an animal (perhaps a cow),
syringing melted wax
into it. He then removed
the brain tissue to produce
an exact cast of the
inner cavities of the
animal's brain.

Leonardo often addresses the notion of the "soul" in his anatomical studies. He studied the soul not only as a vital force, but also from a psychological point of view—what in Aristotelian and Medieval psychology was called the "common sense," or the sensitive faculty in which all external sensations converged. Leonardo tends to present the common sense as the intellective soul, and he made efforts to locate its anatomical seat. In his studies of the skull (in the Windsor Royal Collection) the seat of the soul is located at the crossing points of the intersecting lines inside the skull. In one such study, one of these two lines lies at a distance from the face that measures "one third of the face of man." In other words, Leonardo reversed, inside the body, one of the three divisions visible in the face of the *Vitruvian Man*. The organic and anatomical seat of the soul is, of course, the brain, and Leonardo tries here to give the soul a central position within the skull and the brain.

**RELATED RESEARCH**
See also
ANTHROPOMETRY
page 150

**3-SECOND BIOGRAPHIES**
ARISTOTLE
384–322 BCE
Greek philosopher, he studied
every field of philosophy and
science and was among the
first to study the anatomy of
man and animals

AELIUS GALENUS
ca. 129–200 AD
Prominent Roman surgeon,
anatomist, and prolific of
medical treatises.

**30-SECOND TEXT**
Domenico Laurenza

*The anatomical studies and dissections that Leonardo undertook may have been as much impelled by his search for the soul as a desire to gain knowledge of the workings of the human body.*

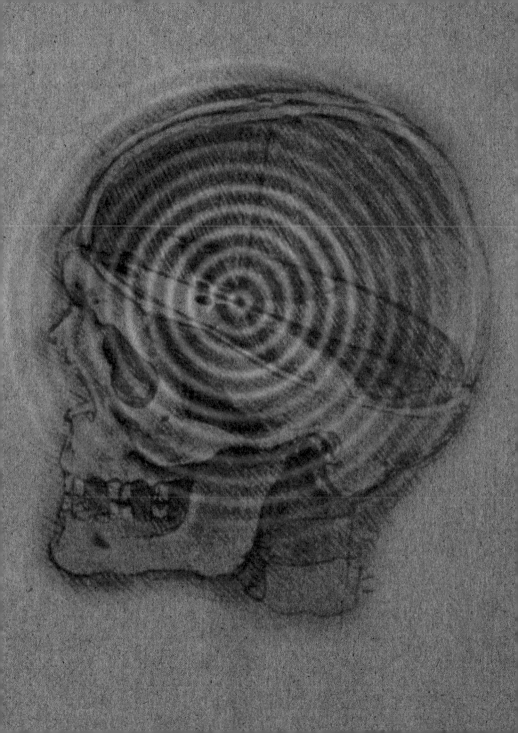

# MAN & ANIMALS

## the 30-second theory

### Leonardo's approach to anatomy

was to emphasize the features that man and animals have in common. In this sense, he defined man starting from "below"—from the animal world. For centuries before and after Leonardo, the main trend in biology would be in the opposite direction and in the 18th century Carl Linnaeus would still classify animals starting from "above," that is from human anatomy and characterizing animals according to what they lack in comparison to man. Leonardo undertook intense studies of comparative anatomy, but unfortunately only a few of them survive. In one example, he compares the legs of a man and a horse, while in another the common aspects of the two are emphasized by putting the man on tiptoe. Emotions, an aspect of the so-called sensitive soul shared by man and animals according to the psychology of that time, are studied in a drawing in which he represents the heads of a man, a horse, and a lion all deformed by an expression of anger and fury. His physiognomic studies of the choleric human type are also closely linked to both comparative anatomy and the study of emotions in man and animals.

**3-SECOND SKETCH**
Leonardo undertook studies of comparative anatomy and of physiognomy in which anatomical and psychological aspects shared by man and animals were emphasized.

**3-MINUTE MASTERPIECE**
Physiognomy judges the character from facial features and was an integral part of Aristotelian zoology and of ancient psychosomatic medicine. Physiognomy studied the permanent character of a person and not momentary emotions. After its rediscovery in the Middle Ages and in the Renaissance, physiognomy enjoyed a revival, shortly before Charles Darwin revolutionized the study of man and animals, through writing The Expression of the Emotions in Man and Animals.

**RELATED RESEARCH**
See also
HELICOPTER
page 58

THE BATTLE OF ANGHIARI
page 88

**3-SECOND BIOGRAPH**
CARL LINNAEUS
1707–1778
Swedish botanist, zoologis and physician.

CHARLES DARWIN
1809–1882
English naturalist.

**30-SECOND TEXT**
Domenico Laurenza

*Entirely against the culture of the day, Leonardo questione the superiority of humans by comparir the anatomical and emotional aspects common to both man and beast.*

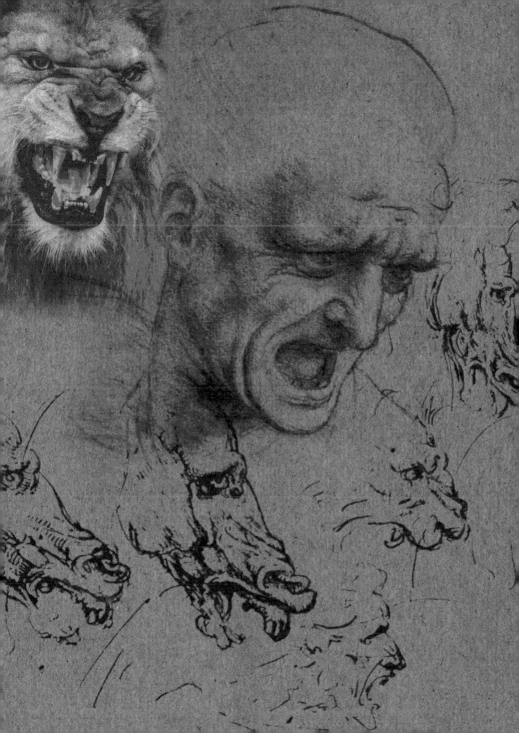

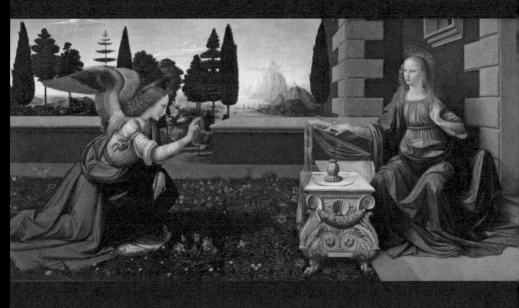

# ANNUNCIATION

The *Annunciation* (in the Uffizi, in Florence) is the earliest surviving independent panel painting by Leonardo. It was probably painted for the monastery of San Bartolomeo a Monte Oliveto, outside Florence, and it could have served as Leonardo's "masterpiece" for matriculating into the painters' guild. Its qualities, and its faults and inconsistencies, suggest that Leonardo painted it sometime between 1472 and 1475, when he was still an assistant in Verrocchio's workshop.

Leonardo had difficulty over the positioning of the Virgin, who is seated too far behind the lectern for her right hand to reach forward, as it appears to do, to hold her book open and point to the page that she reads. The decoration of the lectern, derived from Verrocchio's ornamental vocabulary, is distractingly ornate. The thick, lifeless draperies, which are unnaturally complex, being based on laboriously crafted workshop drawings made with the brush on linen, also signal Leonardo's artistic immaturity. However, the carefully constructed spatial design shows that he was already skilled in geometrical perspective: the orthogonals recede accurately toward a vanishing point at the horizon exactly midway across the panel.

Also anticipating the atmospheric qualities of Leonardo's later landscapes is the muted, hazy tonality of the background harbor set before distant, misty mountains. This backdrop contrasts with the bright colors and verdant richness of the Virgin's symbolic garden—a contrast highlighted by recent cleaning of the panel. Here, his arrival perturbing the Virgin, the Angel Gabriel is about to come to rest kneeling on his right knee, but with his left leg bringing his forward movement to a halt. The lily, symbol of Mary's virginity, that Gabriel holds close to his face is rendered with a precocious naturalism, anticipating Leonardo's subtle observation and recording of nature in plant studies later in his career. Already evidently concerned with narrative communication, Leonardo has ensured that Gabriel's profile, with mouth slightly open in greeting, and his right hand held in a gesture of blessing, are set against dark-toned middle-ground features for greater clarity. The figures' expressive hand gestures, the delicate beauty of their faces, and the rippling curls of their hair farther anticipate qualities of Leonardo's later works. The *Annunciation* is important as an early work by an immensely promising painter, in which many of the qualities of the mature Leonardo are foreshadowed.

*Francis Ames-Lewis*

# ANTHROPOMETRY

## the 30-second theory

During the Renaissance, the proportions of the human body were studied by artists not only to achieve an accurate representation of the human figure in painting or sculpture, but to create an architecture in which the proportions of the different parts of a building correspond to the proportions of the different parts of the human body. In his most famous anthropometric study, the so-called *Vitruvian Man*, Leonardo visually represents the theory of the ancient Roman architect Vitruvius, according to which the human body of perfect proportions fits within a square and a circle. Leonardo's contributions to this field were innovative in different ways. He extended the study of proportions to movement and to time. Indeed, he studied how the proportions of the body vary in the movement from one position to another, for example, from standing to sitting down to kneeling; and how they vary through the different ages of life, as in one drawing in which he compared the measurements of the face of a young and an old man. He extended the study of proportions to include animals, such as horses and dogs. He also studied the proportions of the internal anatomy of the human body, whereas this had previously been limited to the surface of the body.

**RELATED RESEARCH**
See also
SQUARING THE CIRCLE
page 48

MAN & ANIMALS
page 146

**3-SECOND SKETCH**
Leonardo revolutionized Renaissance anthropometry by the inclusion of time and movement, and by its extension to animals and to the inner anatomy of man.

**3-MINUTE MASTERPIECE**
In Francesco di Giorgio's treatises, the human body is inscribed within the plan of a church, with the head corresponding to the apse, the arms to the transepts, and the rest of the body to the nave. In other cases, the human body is inscribed in a column and the face in an entablature. In all cases, the aim is to base the proportions between the parts of architecture on the human body.

**3-SECOND BIOGRAPHIES**
MARCOS VITRUVIUS POLLIO
ca. 80/70 BCE–ca. 15 BCE
Ancient Roman author, architect, and engineer.

FRANCESCO DI GIORGIO MARTINI
ca. 1439–1502
Sienese architect, engineer, and artist.

**30-SECOND TEXT**
Domenico Laurenza

*Leonardo was one of a handful of Renaissance artist-anatomists to give anthropometry a scientific basis on which modern architecture, design, and ergonomics rely.*

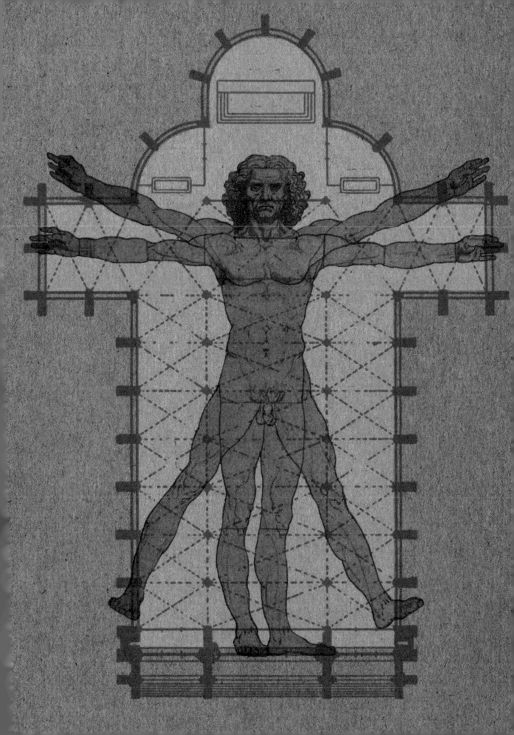

# THE ORIGIN OF LIFE

## the 30-second theory

In an early study, Leonardo represented a cross section of the bodies of a man and woman in the act of intercourse, showing their reproductive systems. Near the representation of coitus, Leonardo writes: "I reveal to men the origin of the first or perhaps second cause of their life." Rather than being an allusion to the traditional theory of the double origin of semen, as is usually speculated, the two causes of life may just be the sperm and the dynamic action of coitus, represented in the drawing. Notwithstanding anatomical mistakes, his drawing is an impressive representation of sex as a dynamic act and an integral part of the reproductive process. In studies begun later in Milan and continued in Rome, Leonardo researched the growth of the fetus in the uterus of a woman, based on both human and animal dissections. Embryology, the study of the period in which the fetus grows and acquires life, led anatomists to speculate upon the problem of the soul. Leonardo was no exception. While studying the relation between mother and fetus, he considered whether the fetus has its own soul and life or whether it is under the tutelage of the mother's soul. Considering such philosophical problems was, at the time, dangerous.

**RELATED RESEARCH**
See also
DISSECTION & VISUALIZATION
page 138

**3-SECOND SKETCH**
Leonardo's anatomical studies of the origins of life include the relationship of the fetus with the mother's body, which might have caused some to seek to blacken his name.

**3-MINUTE MASTERPIECE**
While Leonardo was in Rome (ca. 1513–16), Leo X's Papal Bull *Apostolici regiminis* was issued: the Church regarded the rules to be followed by the natural philosophers (which included anatomists) when they considered the nature and origin of the soul and its relation with the body. A consequence of the Bull was the condemnation and public burning of the treatise *On the immortality of the soul* by Pietro Pomponazzi.

**3-SECOND BIOGRAPHIES**
PIETRO POMPONAZZI
1462–1525
Italian philosopher.

GIOVANNI DE' MEDICI
1475–1521
The son of Lorenzo the Magnificent; became pope with the name of Leo X.

**30-SECOND TEXT**
Domenico Laurenza

*The first artist to depict a fetus in utero, Leonardo's anatomical sketches were startling in their accuracy; had they been made public they could have revolutionized medical knowledge during his lifetime.*

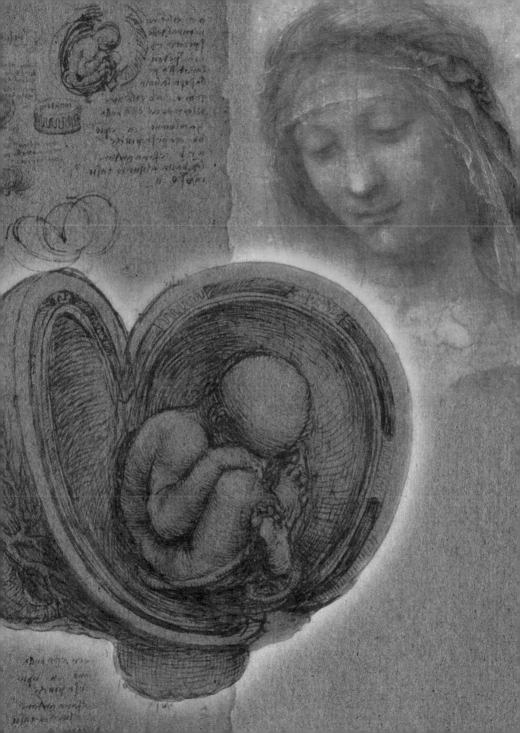

# RESOURCES

## BOOKS

*Geometry and the Visual Arts*
Daniel Pedoe
(Dover, 1976)

*The Golden Ratio*
Mario Livio
(Broadway Books, 2002)

*Le Armi e Le Macchine da Guerra: Il De re Militari di Leonardo, Disegni di Leonardo dal Codice Atlantico (Weapons and Machines of War: Leonardo's De re Militari, Drawings by Leonardo from the Codex Atlanticus),*
Matthew Landrus
(De Agostini, 2010)

*Leonardo da Vinci: Anatomist*
Martin Clayton and Ronald Philo
(Royal Collection Publications, 2012)

*Leonardo da Vinci: Experience, Experiment and Design*
Martin Kemp
(Princeton University Press, 2006)

*Leonardo da Vinci's Giant Crossbow*
Matthew Landrus
(Springer, 2010)

*Leonardo da Vinci: The Marvellous Works of Nature and Man*
Martin Kemp
(Oxford University Press, 2006)

*Leonardo: L'anatomia*
Domenico Laurenza
(Giunti, 2009)

*Leonardo's Machines: da Vinci's Inventions Revealed*
Domenico Laurenza, Mario Taddei, and Eduardo Zanon
(David & Charles, 2006)

*Leonardo on Flight*
Domenico Laurenza
(The Johns Hopkins University Press, 200

*Leonardo, the Inventor*
Ludwig Heydenreich, Bern Dibner,
and Ladislao Reti
(Hutchinson, 1981)

*The Mind of Leonardo: The Universal
Genius at Work*
Paolo Galluzzi (ed.)
(Giunti, 2006)

*Renaissance Engineers: From
Brunelleschi to Leonardo da Vinci*
Paolo Galluzzi
(Giunti, 1996)

## ARTICLES

Landrus, M., "The Proportions of
Leonardo's *Last Supper.*" *Raccolta
Vinciana* 32 (December 2007), pp. 43–100.

## WEB SITES

BBC Science: Leonardo
www.bbc.co.uk/science/leonardo

Digital Archive of Renaissance Manuscripts
www.bibliotecaleonardiana.it/bbl/home.
shtml

Leonardo Bridge Project
www.leonardobridgeproject.org

Leonardo's Machines
www.channel4.com/programmes/
leonardos-machines

The Municipal Library of the Works of
Leonardo da Vinci
www.leonardodigitale.com

The Univeral Leonardo
www.universalleonardo.org

The University of Virginia Digital Archive
www.treatiseonpainting.org

# NOTES ON CONTRIBUTORS

**Francis Ames-Lewis** taught for thirty-six years at Birkbeck College, University of London, UK, and was Pevsner Professor of the History of Art. He has written several books, including *Isabella and Leonardo: The Artistic Relationship between Isabella d'Este and Leonardo da Vinci* (2012) and numerous papers and articles, among which are several on aspects of Leonardo da Vinci's artistic achievement.

**Juliana Barone**, Associate Research Fellow in the School of History of Art, Film and Visual Media at Birkbeck, University of London, UK, has published extensively on Leonardo da Vinci and his *Treatise on Painting*. Her publications also include *Leonardo: Studies of Motion from the Codex Atlanticus* (2011); *I disegni di Leonardo da Vinci e della sua cerchia. Collezioni in Gran Bretagna. Edizione nazionale dei manoscritti e dei disegni* (with M. Kemp, 2010); and *Leonardo: the Codex Arundel* (2008).

**Paul Calter** is Professor of Mathematics Emeritus at Vermont Technical College and Visiting Scholar at Dartmouth College. A graduate of the engineering school of The Cooper Union, New York, he received his M.S. in mechanical engineering from Columbia University, and an MFA in sculpture from the Vermont College of Fine Arts. He has taught mathematics for over twenty-five years and is the author of several mathematics textbooks, including *Schaum's Outline of Technical Mathematics*, *Mathematics for Computer Technology*, and *Technical Calculus*. Calter is involved in the Mathematics Across the Curriculum movement, and has developed and taught a course called Geometry in Art and Architecture at Dartmouth College, which led to his book *Squaring the Circle: Geometry in Art and Architecture*, all under an NSF grant. He is a working artist whose paintings and sculpture often contain geometric and astronomical themes.

**Brian Clegg** (www.brianclegg.net) has a natural sciences degree from Cambridge and has worked for British Airways and his own creativity consultancy. He has written for a wide range of publications from *The Observer* to *Playboy*, and has 18 published popular science books, most recently *Dice World*.

**r. Matthew Landrus** is a Research Fellow at Wolfson College and the History Faculty at the University of Oxford. He has published books and articles on Leonardo, including *Leonardo a Vinci's Giant Crossbow* (2010), *Le Armi le Macchine da Guerra: Il De re Militari Leonardo* (2010), and *The Treasures of Leonardo* (2006).

**Domenico Laurenza** is a science historian, an expert on Leonardo da Vinci's scientific work and on the history of anatomy and technology in the Renaissance. He is now devoting a substantial portion of his attention to the history of geology. He is scientific consultant of bgC3 (Seattle-Kirckland) and Museo Galileo (Florence). He has taught in several universities worldwide, including the University of Florence and McGill University, Montreal, and has been fellow of several scientific institutions, including the Warburg Institute in London, the Metropolitan Museum of Art in New York, and the Italian Academy at Columbia University. He is the author of many books; available in English are: *Leonardo on Flight* (2007), *Leonardo's Machines: Da Vinci's Inventions Revealed* (2006), and, most recently, *Art and Anatomy in Renaissance Italy: Images from a Scientific Revolution* (2012).

**Marina Wallace** is Professor of Curation and Director of Artakt, Central Saint Martins College of Art and Design, University of the Arts, London. She has a background in classics, fine art, art history, and journalism. She curated a number of groundbreaking exhibitions including "Seduced Art and Sex from Antiquity to Now" (2007/8) and "Spectacular Bodies: The Art and Science of the Human Body from Leonardo to Now" (2000/1). She is the author of a number of publications, amongst them *John Hilliard, 1969–1996* (1999), *Spectacular Bodies: The Art and Science of the Human Body from Leonardo to Now* (2000), *Head On, Art with the Brain in Mind* (2003), *Mendel, the Genius of Genetics* (2003), *Seduced, Art and Sex from Antiquity to Now* (2007), *Acts of Seeing,* (2009), *La Cultura Italiana,* Volume X (2010), and *The Lives of Paintings: Seven Masterpieces by Leonardo da Vinci* (2011).

# INDEX

# ACKNOWLEDGMENTS

PICTURE CREDITS
The publisher would like to thank the following individuals and organizations for their kind permission to reproduce the images in this book. Every effort has been made to acknowledge the pictures; however, we apologize if there are any unintentional omissions.

All images from Dover Images/www.doverpublications.com and Shutterstock, Inc./www.shutterstock.com unless stated.

**AKG Images**/De Agostini Picture Library: 67, 91T, 99.

**Biblioteca Leonardiana**, Vinci, Italy (Leonardo da Vinci, Traité de la peinture, Paris, Langlois, 1651, p. 67): 17.

**Corbis**/Alinari Archives: 79BR, 85, 87, 88, 129; Baldwin H. Ward & Kathryn C. Ward: 63B; PoodlesRock: 131.

**Nevit Dilmen**: 63TL.

**Timur Kulgarin**:19B.

**Erik Möller** (Leonardo da Vinci. Mensch - Erfinder - Genie exhibit, Berlin 2005): 91B.

**Scala Archives**: 122; British Library board/Robana: 71; Courtesy of the Ministero Beni e Att. Culturali: 148; Veneranda Biblioteca Ambrosiana/DeAgostini Picture Library: 69, 113.

**Topfoto**/Print Collector: 47, 81.

**Luc Viatour**/www.Lucnix.be: 49.

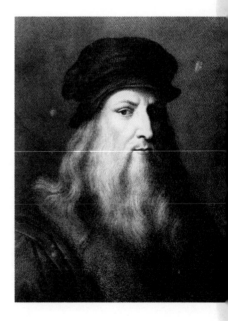